The Woman Lawyer

making the difference

To my parents, Archie and Leah,
and to Ian

The Woman Lawyer
making the difference

Clare McGlynn   BA, M Jur, Solicitor
University of Newcastle upon Tyne

Butterworths
London, Edinburgh, Dublin
1998

United Kingdom	Butterworths, a Division of Reed Elsevier (UK) Ltd, Halsbury House, 35 Chancery Lane, LONDON WC2A 1EL and 4 Hill Street, EDINBURGH EH2 3JZ
Australia	Butterworths, a Division of Reed International Books Australia Pty Ltd, CHATSWOOD, New South Wales
Canada	Butterworths Canada Ltd, MARKHAM, Ontario
Hong Kong	Butterworths Asia (Hong Kong), HONG KONG
India	Butterworths India, NEW DELHI
Ireland	Butterworth (Ireland) Ltd, DUBLIN
Malaysia	Malayan Law Journal Sdn Bhd, KUALA LUMPUR
New Zealand	Butterworths of New Zealand Ltd, WELLINGTON
Singapore	Butterworths Asia, SINGAPORE
South Africa	Butterworths Publishers (Pty) Ltd, DURBAN
USA	Lexis Law Publishing, CHARLOTTESVILLE, Virginia

A CIP Catalogue record for this book is available from the British Library.

ISBN 0 406 89593 7

Printed by Redwood Books Ltd, Trowbridge, Wilts

Visit us at our website: http://www.butterworths.co.uk

Foreword

'Law! But why law.' To my mother, it was unfathomable. Why would anyone choose to study anything so alien and, worse still, choose to do it in the hell that is London? Higher education was itself unknown and forbidding territory; law was a snare to be avoided by the working classes; but the far reaches of the unspeakable South were beyond the pale.

However, to me the law and London combined the lure of the exotic and I liked the surprise it engendered when people in Glasgow heard what I intended to do. The warnings followed. It is a world full of toffs. You have no contacts. It is no career for a woman. Being away from home is too expensive for working class people; at home, a student is just one more mouth to feed. English law will be no good to you in Scotland. You'll never survive.

For my family, my study of law felt like a very large departure from my roots. I, in my turn, rationalised that my ambitions were not grandiose. I would use a law qualification as a step into social work or I would work for a trade union. I wanted to do some good and I knew from my own parents' experiences that ordinary people were not empowered by the law but in dread of it. It meant frightening expense; it meant contact with lawyers who knew little about the reality of their lives. The law was not theirs; it belonged to other people. With all the simple belief of youth, I thought I could make a difference.

The fears and reservations expressed by my family and friends seeded my determination to make a go of it but over the years I saw clearly why they felt they had no purchase on the legal system and why it had to change.

In my day, it was possible to study directly for the Bar at the Inns of Court School of Law without doing an initial degree. The course meant you qualified as a Barrister-at-Law and that was what I did. Oblivious to the subtle distinctions of the English education system, I had no idea that this was considered a somewhat inferior way of becoming a lawyer. I just thought quicker was better.

At first I assumed that all these young men were smarter than I, especially since they spoke in such cultured, radio-announcer tones. There was only one other woman in my tutorial group. I was reticent about contributing to discussions, sure I would make a fool myself. I became watchful, realising slowly that what these men had was self-assurance rather than great intellect. They were confident of their place in the world; barristers in the making. Even in their student scruffiness, I could envisage them in the future, thumbs under their lapels, holding forth in the courts of law. How was I ever going to be able to do it?

I settled on my grandmother's unattractive adage that there was more than one way to skin a cat. When hurdles presented themselves I either staggered over them or looked for a way round them. In fact, I did what women have always done. I gritted my teeth and got on with it.

I qualified in 1972 and embarked upon pupillage with a dear man who took me on unwillingly, after he was worn down by my endless entreaties. His warnings about the misogyny in his chambers were all true and I had six months of invisibility, totally ignored. My second six months was more fulfilling, as I moved to another set of chambers where I started doing some casework for the newly fledged law centres. I was paid almost nothing but had got used to living hand-to-mouth as a student. I loved the work which was precisely what I felt I had come to the law to do, acting for people who had little voice within the system.

When I could not find a tenancy, I set up a new set of chambers with a group of other young barristers: three men and three women. From there, my practice developed from low level crime and civil liberties work through to the most serious crime and human rights cases, which I now conduct.

The story is not without its hiccups. You cannot criticise discrimination with the law and the legal profession without the beneficiaries of that bias taking umbrage. And some of them really did take umbrage. You cannot point to the law's failures without the establishment closing ranks, at least in the beginning. I am still bemused when I am described as 'that woman' by some of the older judges.

Yet my real dismay was when senior women were not supportive of their own gender. Even now there are privileged women who are unconscious of the disadvantage experienced by other women; money and family connection can be a great cushion against potential discrimination. Others who are successful listen to the siren call that, having succeeded in a male world, they are exceptional and they bask in the belief that they are truly blessed amongst women.

Although I have had forays into other pastures, including broadcasting and writing, nothing has replaced my passion for the law. No two cases are the same and the challenges are constantly renewed. It means that I never discourage anyone from joining because I love it so much and I think those who are most often diverted are those whom the law most needs.

In the years since I started practising law, the profession's landscape has radically changed. One of the most heartening shifts is that women from so many different backgrounds are coming into the law, bringing with them fresh perspectives. Black women, white women, women of all ages. The tutorial class with only a trickle of women has transformed and in most law schools half the students are women. Women are coming into chambers and law firms in increasing numbers, and the law faculties include a growing cohort of women academics. These women do not disregard their background but draw upon it in their work, and the richness of that experience is humanising the law.

The debate about women's inclusion is no longer completely marginalised; the professional bodies proclaim their anti-discrimination policies and the Lord Chancellor declares his commitment to equal opportunities. With increasing frequency women are being appointed to the Bench. 'You have come a long way, baby', they tell us, as though the chapter is almost closed.

Yet many women still find the odds stacked against them in ways that are not experienced by men. More women use their law degree to enter other careers because pupillage and traineeships are so hard to obtain. Certain highly remunerative areas like commercial law, revenue law, patents and intellectual property are almost exclusively male zones. In a proudly adversarial process, the combative advocate is lauded except when female. Senior partnerships in solicitors' firms are still disproportionately offered to male contenders and the numbers of women in silk or in judicial roles are still shamefully low. Many women leave practice when faced with the balancing act of their law career and family life, because they feel under pressure to perform in the same competitive way that their male colleagues do.

Clare McGlynn has produced an absorbing and illuminating study of women in the law, charting their journey, documenting attitudes and hurdles, describing the changes which are necessary before women will play an equal role. Her book is a testimony to women's determination to lay their claim upon the law and contribute to one of society's central institutions in a meaningful way. The law regulates social relations and, as the role of women in society has changed, it has become the more crucial that women contribute to legal discourse and the interpretation of laws.

This book is essential reading for every woman in the law. It is our narrative. Each and every one of us has a contribution to make as the story unfolds. The law needs women because women need the law.

Baroness Helena Kennedy QC
Doughty Street Chambers
London WC1

Acknowledgements

In the planning and writing of this book, I have benefited enormously from the advice, assistance and support of a number of people. Without Nina Hall, the idea for the book may never have come about. It was during one of our many discussions on women, feminism and the law that we decided there was a clear need for a book such as this. I would like to thank Nina for her insight, tact, wisdom and friendship which have proved invaluable in the development and drafting of this book.

I appreciate the support of Newcastle Law School, particularly in providing me with the research assistance of Clare Maher, who worked well beyond the call of duty (and pay!). Clare's research skills, initiative, ideas and commitment to her work, and the book, were of immeasurable assistance. Catherine Farrelly helped with the checking of proofs. A number of colleagues have also helped enormously. I highly value Richard Collier's constant advice and encouragement. Thanks are also due to Joanna Gray and Chris Riley, and, in particular, to Alison Dunn for considering with me many of the ideas discussed in this book, as well as helpfully commenting on draft chapters. The views expressed have also benefited from discussions with many other colleagues, albeit often unintentionally on their part.

There are many colleagues and friends throughout the academic community who have helped in the formulation of my thoughts and ideas, as well as in the provision of useful information, including Rosemary Auchmuty, Fareda Banda, Jayne Barnard, Fiona Boyle, Tammy Hervey, Katherine O'Donovan, Richard Moorhead, Fiona Raitt, Hilary Sommerlad and Pat Twomey. I would also like to thank the various heads of department and other academics who took the time to complete the questionnaire which is considered in Chapters 1 and 2.

It is not only fellow academics who have helped inspire and support this project, but also many of the women law students with whom I have had the pleasure of working. Their ambition, determination, energy and ability never ceases to amaze and impress me. Talking with them has also ensured that I do not forget, as is easier the more senior one gets, that issues facing women in the law are not just concerned with promotion, or reconciling work and family life, but with actually getting a job and studying and training in an environment free from harassment and discrimination.

My time in practice and involvement with the group Young Women Lawyers has brought me into contact with many inspirational women lawyers who have over the years helped shape my views. Thanks particularly to Gill Baxter, Jill Brown, Rebecca del Tufo, Nancy Duffield, Catherine Fitt, Caroline Graham, Marilyn Gulland, Barbara Hewson, Amanda Rooney, Tilly Rubens, Samantha Steer, Rachel Tetzlaff-Sarfas, Liz Veats and Judith Willis. Many institutions and organisations have provided much needed information and advice, and I would like, in particular, to thank Marcia Williams of the Law Society, Fiona Waye of CUCO, as well as the Bar Council, the AUT and HESA.

A particular thanks must go to all the contributors to the book, both for their personal testimonies, and for their perseverance with the editorial process. Unfortunately, due to space, I was not able to include every contribution. My thanks goes to those who took the time to prepare some text for the book, but which in the end, could not be included. The support of Baroness Helena Kennedy QC in writing the foreword for the book is also greatly appreciated.

Had my parents not persuaded me that completing my solicitors' training would be worthwhile, this book would not have been possible or more likely ever imagined: they did indeed know better! They have also proved a tower of strength throughout the writing of this book. My father has been a careful and insightful reader of the entire text of the book; his perception, vision and attention to detail proving invaluable. I hope too that some of my mother's common sense does, perhaps, shine through the text.

The final words of thanks must go to Ian Ward. Ian has sustained me through the life of this and many other adventures. He has read and commented on each of the chapters of the book, but, moreover, has debated with me, on an almost daily basis towards the final stages of drafting, the ideas, controversies, nuances and progress of the book. I am indebted to Ian for his patience, care, humour and conversation.

Contents

Abbreviations

ACLEC	The Lord Chancellor's Advisory Committee on Legal Education and Conduct
ALRC	Australian Law Reform Commission
ALT	Association of Law Teachers
AUT	Association of University Teachers
AWB	Association of Women Barristers
AWS	Association of Women Solicitors
BA	Bachelor of Arts
BCL	Bachelor of Civil Law
BVC	Bar Vocational Course
CACH	Centralised Applications and Clearing House
CHULS	Committee of Heads of University Law Schools
CRE	Commission for Racial Equality
CVCP	Committee of Vice Chancellors and Principals
CUCO	Commission on University Career Opportunity
EOC	Equal Opportunities Commission
FILEX	Fellow of the Institute of Legal Executives
ILEX	Institute of Legal Executives
HEFCE	Higher Education Funding Council for England
HESA	Higher Education Statistics Agency
LCD	Lord Chancellor's Department
JAC	Judicial Appointments Commission
LLB	Bachelor of Laws
LPC	Legal Practice Course
LSF	Law Society Final Examinations
NATFE	National Association of Teachers in Further Education
OU	Open University
PACH	Pupillage Applications Clearing House
RAE	Research Assessment Exercise
SBL	Society of Black Lawyers
SPTL	Society of Public Teachers of Law

List of tables

Introduction:
'From naïve and passive to aware and active'

'Only when women are aware of the extent of the discrimination against them, of how it operates and of how to use the law and to influence law reform to their own ends, will further progress be made.'

Susan Atkins and Brenda Hoggett, *Women and the Law*, 1984[1]

If you are a law student, legal academic, woman in legal practice, policy maker in the law, or seeking to explore equality and fairness in our society, this book is for you. The remit of my book is, therefore, unashamedly broad, for my purpose is to speak beyond the confines of the few and to encourage a more informed public debate. What I have set out to offer is a new analysis of the role and status of women in the law, from the law school to the judiciary. In doing so, I have drawn on the recent literature and research, much of which is not particularly accessible, which has been subject to little rigorous analysis and which is rarely placed in its broader theoretical and political context. Moreover, empirical research carried out specially for the book has enhanced my analysis by revealing new insights into the representation of women in law schools and the extent to which gender perspectives are included in the law school curriculum.

In the pages ahead, I go on to demonstrate the continuity of exclusion and marginalisation of women in the law, from law school to the judiciary. Underpinning the analysis is the view that progress is not simply a matter of women seeking to try harder (to be like men). There are real structural and institutional disadvantages and discriminations facing women which require more than individual reformation to overcome, although it is not my intention to suggest that there is no agency required on the part of women.

It is one thing to document the increasing representation of women, together with analysing the obstacles in the way of women, and quite another to offer explanations as to why those obstacles exist. This book concentrates on the documentation and analytical tasks, as no work in the

UK has yet made such an attempt. I leave the latter task to others and future work, work which I hope will be enhanced by the analysis offered in this book. While I do not dwell on the theoretical debates examining the status of women in the law, I acknowledge the depth of their contribution in providing the inspiration for this study.

To this end, I hope to have written a book which I wanted to read. Throughout my time as a law student, teaching fellow, trainee solicitor and law lecturer, finding information and scholarship on women in the law has proven surprisingly difficult. Sure enough, there is a growing body of feminist literature examining the ways in which the substantive law impacts on and constructs women: but very little on women working and studying *in* the law.

Things have been changing, and indeed, in the last few years the under-representation of women in the legal profession has received a much higher public profile. It might even be said that it has become an 'establishment' issue. Successive research studies have been widely covered in the press (especially if they concern sexual harassment), and 1995 saw the introduction of an annual Woman Lawyer conference at which senior members of the profession speak; this year being no exception with a lecture from the Lord Chancellor. However, the extent to which there is real understanding about the role and status of women in the legal profession because of such conferences and press commentary is not clear. Reports, for example, produced by both the Law Society and Bar Council on the situation of women, are not widely available. It is encouraging to see articles appearing in the press, but this still only gives a patchwork coverage, and is necessarily confined to 'news', as opposed to rigorous analysis. Crucially, although understanding regarding women in the legal profession may be growing, few, if any, links are made with law schools and the academic study of law. This failure, along with the absence of research studies of the experiences of women law students or women legal academics, militates against an informed debate in the UK and leaves us trailing behind North America and Australia.

As the book embraces a broad audience, the objective is not just to analyse and raise the level of public debate, but to inform, encourage and empower women studying and working in the law. I hope to speed up a process which Professor Gillian Morris-Kay has described as her 'painful transition' from being 'naïve and passive to aware and active' regarding the inequalities which women still face in the workplace.[2] Part of this educative process is to engender an awareness of the collective nature of both the problems and the solutions. For example, much of the writing and advice which is available to women lawyers is individualised and targetted at what *women* can do to improve and change themselves. This reaches absurd heights when two national newspapers publish survey results, with illustrative diagrams, showing how women lawyers should dress to achieve success.[3] Undue concentration on the personal attributes of women diverts attention away from the vital issues of occupational and professional structures, ideologies and recruitment patterns which effectively discriminate against women.[4]

Similarly, women often talk about seeking 'special arrangements' from their firm to deal with, for example, maternity leave, or of gaining a 'good deal' from their chambers or employer in terms of seeking appropriate working conditions. Rarely do women connect these common concerns and struggles to the economic structure of the firm and the legal profession, to the nature of the law and legal culture, or to the fact that women as a whole are disadvantaged in society as well as in the legal academy and profession.[5] What I am arguing here is that change and reform should be seen as an issue for all women, not just for individual women to try to work out on their own: connections between women must be made. Mona Harrington has argued that for women 'to work professionally to promote the general equality of women, without recognising clearly the remaining force of cultural oppressions, is self-defeating'.[6] Baroness Helena Kennedy QC is right: 'women have gone through the stage when they did the adjusting; now it is time for the institutions to change'.[7]

Thus, my book is premised on the belief that knowledge really is power, and women need power if they are to achieve success in the law. There is little point in writing a book which does not deal with the gendered nature of the law school and legal practice. To do so would only tell a partial story and would not equip women for the challenges which could face them; nor would it empower women to take action demanding a change to academic and workplace cultures and traditions. In would also propagate a myth that the legal academy and profession are open to all, free from discrimination.

The dilemma, therefore, in writing a book such as this is that in recognising the remaining force of cultural oppression in the law, there is a possibility of demoralising and discouraging women. However, the driving force behind this book is not just to inform, but to motivate and inspire women. It is very much my hope that the inspiration will, in part, come from the personal testimonies of the many women law students, academics, practitioners and judges which enrich the text. Given that there are so few women role models in the law, the personal testimonies offer a fascinating insight into the diversity of experiences and approaches of women lawyers. Similarly, at a time when the traditional (masculine) work pattern remains very dominant in the law, it is my view that the personal testimonies can help to validate a diversity of career options, as well as giving voice to the choices made by women to negotiate work and family commitments.

In choosing the contributors an attempt was made to select a cross-section of women in terms of, for example, their age, status, work experience, ethnicity and sexual orientation. However, the contributors are not presented as representative of all women in the law. Women's experiences are clearly mediated through their sex, nationality, ethnicity, sexual orientation, class, age, parenthood and many other factors. Just as there is a diversity of women, there is a diversity of women in the law.

The choice of contributors was made from those known personally to the author, or through colleagues, and those known publicly for their work. Selection was not made on the basis of whether their views on women in the legal profession and/or feminism were 'known' or 'right'. Indeed, the strength of the contributions comes from the richness, openness and diversity of views expressed and perspectives adopted. Each contributor approved the use of their contribution in the book, and a full list of contributors can be found in appendix 7. The contributors were given a framework within which to relate their personal histories and views, with suggested topics including background, inspirations for choice of legal career, experiences as a woman in the law, advice for other women lawyers, examples of good practice and views on the future. However, it was a conscious decision that no editorial direction was given, although it is accepted that in writing a contribution for a book on women in the law, a particular perspective was already taken for granted. The contributions were used in two ways; first, excerpts have been integrated into the text of the book, and secondly, the full text of a number of contributions has been included at the end of relevant chapters.

Choosing women as the subject of study has many problems. For example, Regina Graycar has argued that each time we use the prefix 'woman' before the term judge, lawyer or professor, we reinforce the fact that men are the norm, and women exceptional.[8] Thus, Katherine Bartlett has suggested that 'using difference as a category of analysis can reinforce stereotyped thinking and thus the marginalised status of those within it'.[9] It is my view that if we never say 'woman judge', we will almost always think of a man; if we never talk about 'women professors', there will remain an unspoken image of the professor as a man. If we do not consider the fact that women are disadvantaged, because we fear drawing attention to the fact that women are different from men, we will not be able to start to tackle the causes of such disadvantage.[10] Accordingly, this book is unashamedly about *women* in the law.[11]

One final theme provides my inspiration for this book. The status and role of women in the legal academy and profession is not just a question of ensuring the professional authority and career advancement of women. Undoubtedly women in the law are entitled, as in any other area of work, to progress on their own terms, but the analysis offered, and prescriptions given in this book are seeking more than this. I do believe that a greater number of women legal academics, practitioners and judges will make a difference to the law, academia, the profession, the Bench and public confidence in all four. I say this, not on the basis that women have certain innate characteristics that mark out *all* women from *all* men, but on the basis that in this society many women do have very different experiences to many men and that they bring these to their work. Similarly, I believe that it is not enough for women in law to seek equality for themselves, without making similar demands for women as a whole.

Ultimately, changes wrought by increasing representation of women at the senior levels of the academy, profession and judiciary will be beneficial for women and men, and for all 'outsiders'. As Mary Jane Mossman has argued: 'The challenges to redefine the law in women's interests, and to redefine appropriate roles for women and men as lawyers, are just part of a larger process to make the law more responsive to all members of society, and to make the legal profession accept and value equally the talents and experiences of all women and men in a modern multicultural community.'[12]

NOTES

1 Susan Atkins and Brenda Hoggett, *Women and the Law*, Blackwell: Oxford, 1984 p 5.
2 Quoted in *The Times Higher Educational Supplement*, 26 July 1996.
3 Glenda Cooper, 'Female lawyers told dress holds you back', *The Independent*, 2 March 1998; Clare Longrigg, 'If you want to get ahead then get a suit, women lawyers told', *The Guardian*, 2 March 1998.
4 A point made over fifteen years ago by David Podmore and Anne Spencer, 'Women Lawyers in England', 9 (1982) *Work and Occupations* 337–361, p 338.
5 A point reinforced in the US context by Mona Harrington, *Women Lawyers – Rewriting the Rules*, Plume-Penguin: USA, 1993, p 32.
6 Ibid p 249.
7 Helena Kennedy, *Eve was Framed*, Chatto & Windus: London, 1992, p 263.
8 Regina Graycar, 'The Gender of Judgments: An Introduction', in Margaret Thornton (ed), *Public and Private – Feminist Legal Debates*, Oxford University Press: Oxford, 1995, pp 262–282.
9 Katherine Bartlett, 'Feminist Legal Methods', 103 (1990) *Harvard Law Review* 829–888, p 835.
10 See also Margaret Thornton, *Dissonance and Distrust: Women in the Legal Profession*, Oxford University Press: Oxford, 1996, p 5.
11 However, this is not to suggest that all men occupy a privileged position in the law school or legal practice. Those men who do not conform to the expected masculine norm may also find themselves disadvantaged. Accordingly, it has been argued that we need a greater understanding of what constitutes the 'masculinism' of law and legal culture, and of men as legal actors, if we are to resolve the problem of 'women in the law': see Richard Collier, '"Nutty Professors", "Men in Suits" and "New Entrepreneurs": Corporeality, Subjectivity and Change in the Law School and Legal Practice', 7 (1998) *Social and Legal Studies* 27–53. We do indeed need to

move beyond the cultural forces which valorise the various guises of masculinity, but gaining such understandings is, however, another research project, and one which will be better informed and better able to address reform, if the role and status of women is clearly understood.

12   Mary Jane Mossman, 'Women Lawyers in Twentieth Century Canada: rethinking the image of Portia', in Regina Graycar (ed), *Dissenting Opinions – Feminist Explorations in Law and Society*, Allen & Unwin: Sydney, 1990, p 95.

Chapter 1
Women learning the law

'Mrs Belva Lockwood

The Faculty of Columbian College have considered your request to be admitted to the Law Department of this institution and after due consultation have considered that such admission would not be expedient as it would be likely to distract the attention of the young men.

Respectfully

Geo. Samson, President', 1869[1]

And so in 1869 a putative woman law student was rejected on the basis that she might be a distraction to her male colleagues. This sort of comment was not unusual. It was said by a prominent alumnus of Yale Law School in 1872 that he was in favour of women studying and practising law, but only 'provided they are ugly'.[2] Nonetheless, despite such discouraging beginnings, since the 1970s women have been entering law schools in rapidly increasing numbers and since 1988 there have been slightly more women than men law students in the UK.[3]

This chapter has two principal aims. The first is to provide women considering law school, a legal career, or postgraduate legal studies, as well as the law student, with an interesting and thought-provoking approach to entering law school and pursuing legal studies, and equip her with an understanding of the gender issues which underscore the study of law and the life of a law school. Many of these issues will then be taken up in the following chapter on women legal academics. In addition, the chapter explores the impact of gender on the law school and the legal curriculum, drawing on the insights offered by empirical research carried out for this book. In conclusion, the personal testimonies of two recent law students expand on and illustrate many of the themes which follow.

GETTING TO LAW SCHOOL

Why study law?

Ruth Deech, Principal of St Anne's College Oxford, recalls that it was her father's experiences which inspired her to study law. 'My father was a refugee from Vienna, where he had studied law, but had not been allowed to practice by the Nazis', says Ruth. 'He had enjoyed studying law but had eventually been drawn into journalism as a means of self-defence for him and other persecuted Jews of the period.' As a result of this early experience, Ruth 'formed the notion that only the rule of law and a genuine commitment to it at all levels of society could protect minorities from persecution and give every member of the population equal opportunities and access'.

Barrister Elizabeth Woodcraft was also driven to study law as a means by which to further her civic aims. Elizabeth worked as a national co-ordinator for the Women's Aid Federation, the organisation providing refuges for women escaping violence in the home. She also worked closely with Jo Richardson MP who introduced the Domestic Violence and Matrimonial Proceedings Bill which was a major breakthrough in the protection of women suffering abuse in the home. Elizabeth recalls that having been involved in such areas of work, and having seen the positive role which law can play, she decided to study law and become a barrister.

Rachel Tetzlaff-Sarfas, a partner in the solicitors' firm Thompsons, remembers that she became fascinated with the law in her early teens, first being intrigued by reading sensational reports of criminal trials in her parents' newspapers. 'The principles of justice, fairness and equality', Rachel says, 'have always been important to me. I saw the law as the tool I could use to uphold my beliefs'. Rachel wanted to become not only a 'champion of justice, but also to make the law more accessible'.

But not everyone wishes to study law for such specific reasons. Like many law students, Jane Hyndman, who is a solicitor with the company Live TV!, says that she has 'no clear recollection why I chose to study law except perhaps that it would lead to a good qualification and would be useful even if I ultimately chose a career in a different field'. Natalia Siabkin, a freelance solicitor, wanted to study either English or classics, but in the end made a pragmatic decision to study law 'as a conduit to a career'. Similarly, Fiona Boyle, senior lecturer in law at Northumbria University, 'chose law because it fitted better with my feeling that I wanted a vocation'.

But what is law?

Often the decision to study law is a difficult one to make, sometimes because it is not clear exactly what law is or what the study of it entails.[4] The law impacts on all our lives; it is a rich tapestry of rules, statutes, morals and politics. An Act of Parliament or a judgment in a case are examples of the law, but there is more to the law than to seeing something written down in a case or statute, or trying to find *the* answer to a specific problem (especially as more often than not there is no definitive answer). The law is also about the principles of justice, rights and democracy; about regulating society and shaping the law to change society. As we saw above, many women have decided to study and practise law in order to seek to improve the lives of others. In this way, the law can be seen as a tool to be used to ensure justice, fairness and equality for all.

Elizabeth Woodcraft is particularly proud of one case in which she was involved which exemplifies the complex nature of the law and its interaction with people's lives. Elizabeth acted for a 17-year-old woman who was charged with incest with her own father. For technical reasons, the woman had no legal defence and Elizabeth was told by the judge that she should advise her client to plead guilty and, in return, the judge would give the woman a lenient sentence. Elizabeth responded that her client, despite her having no legal defence, would be pleading not guilty on the basis that she did not deserve a criminal conviction. Elizabeth was accused by the judge of simply using her client to make a political point, as Elizabeth did not like the current law which meant that this woman could be charged with the offence. But, Elizabeth's strategy won in the end, and saved her client the trauma of a trial and a criminal conviction, because a later judge agreed with Elizabeth's position and persuaded the prosecution to drop the case.

What this case shows is that the law is not as simple as the actual words written in a statute, but is intimately bound up in the context of the case, the political and cultural attitudes that prevail at the time, and the personal commitment of lawyers and judges. Barrister Josephine Hayes notes that 'any legal system tends to reflect the prejudices of the rule-makers. Take for example the old common law rules, all judge-made, that women could not vote, serve on juries, or hold office as judges. Further examples are the rules that in rape cases women's uncorroborated evidence could not be trusted and that on marriage their property became that of their husbands, although these provisions have all now been reformed by legislation.'

In many ways, the law regulates the whole of our society from issues of crime to the very economic basis of our economy. Caroline Graham, a City solicitor acting for corporate clients, enjoys engaging with the law as it affects businesses because 'our society, in common with most others in the world, is underpinned by commerce and trade, and as a corporate lawyer you are part of the system which regulates and controls the capitalist economy which is so fundamental to our way of life'.

Law school admissions: ensuring equal opportunities

Having decided to study law, thoughts turn to how to apply to law school.[5] For a full-time degree, applications are made via the Universities and Colleges Admissions Service (UCAS); applications for all other courses being made direct to the institutions concerned. Universities select students by reference to a 'standard offer'[6] and the application form, particularly the personal statement and reference. The interview, although generally employed for part-time courses and mature students, is increasingly rare. Its passing, however, should not be mourned as it is a notoriously inadequate process for deciding whether or not someone is suitable for a place at university, or a job.[7]

Although some universities and law schools have an 'admissions policy', the admissions processes of many universities are shrouded in mystery and the practices which are seen as obvious and appropriate for the recruitment of employees are sometimes lacking. One study of law school admissions policies described them as 'bizarre', with the potential for them to be seen as 'capricious', especially from the point of view of the applicant.[8] In a later study, the criteria of 'enthusiasm' and 'motivation' were rated as very important by the vast majority of admissions tutors, although it is not clear how, if at all, such attributes are or can be measured.[9] Such criteria are highly subjective and prone to the biases and prejudices of individual tutors, which may be consciously or unconsciously discriminatory.

The influential Lord Chancellor's Advisory Committee on Legal Education and Conduct (ACLEC) has expressed concern 'at the absence in some institutions of transparent admissions policies using objective criteria'.[10] What we require, therefore, in all law schools, is a comprehensive admissions policy which is clearly linked to the needs of the law school and the requirements of the degree course, and takes into account the need to comply with equal opportunities.[11] There also needs to be training in equal opportunities for all those involved in the admissions process, as well as evaluation and monitoring of the policy.[12] How many law schools collate information as to the ethnicity, gender, types of secondary school attended, qualifications obtained or occupational status of parents of applicants, of applicants, those offered places and those who accept? Only if statistics are collected and considered over a number of years can universities be sure that they are admitting students on merit alone, as they claim to be.[13] Law schools have a crucial role to play in shaping the legal profession of the future, and taking that responsibility begins with admissions.

Factoring in gender: choosing a law school

There are many factors which will come into play in making a decision about which university to attend. The type of degree favoured will be influential, as will a wish to live close to, or far from, home. The criteria are likely to include the quality of education provided, especially now that every university is graded in terms of its teaching quality and research excellence. The nature of the institution in terms of its facilities, including library, computers and residences will also be important. Many of these and countless other suggestions are included in the numerous books which seek to assist applicants in this difficult decision-making process.[14] However, there are other important factors, such as the nature of the law school curriculum and the numbers of women academics in each law school, to which reference is rarely made and which are therefore examined below.

Giving consideration to these factors has, however, been difficult, owing to the paucity of information available. Accordingly, research was carried out for this book into the integration of gender/feminist issues into the curriculum and the gender composition of each law school. Appendix 2 lists all the universities which offer courses on 'women/gender and the law' (with a few exceptions), and appendix 3 gives detailed information on all university law schools (again with a few exceptions) regarding the extent to which gender is integrated into the curriculum. Although the ability of a law school to offer certain courses may depend on the particular staff in the law school at any one time, the ethos and culture of the department may also play an important role in determining whether there are staff in that law school with interests in gender/feminist approaches to the law, and, if so, whether they are able to offer courses in these fields. Why the integration of these perspectives in the legal curriculum is so important is considered later in this chapter.[15]

A further factor worth considering is the gender composition of the teaching staff in the law school.[16] Appendix 3, and table 2.2 in the following chapter, detail the gender composition of each university law school (with a few exceptions). This data must be read and interpreted with care, and readers are encouraged to explore the results with individual law schools.[17] But what difference does the number of women legal academics make? Increasing numbers of academic women in law schools have begun to make the law school a more diverse place. This has brought new and innovatory challenges to the law, legal culture and legal scholarship. With diversity comes difference in teaching modes, student interaction, culture and ethos, alternative perspectives from which to teach and research, cumulatively engendering a more open and accessible environment. The representation of women in the law school may also have an impact on the experiences of women law students, a possibility which is considered later in this chapter.[18]

Who succeeds? Law school applications and acceptances

In 1996, 10,716 women applied to undertake a law degree in the UK, comprising 58 per cent of applicants and a similar percentage of those accepted (see table 1.1). It can also be seen from table 1.1 that the proportion of women law students to men is increasing year on year. It will be interesting to see, in the years to come, whether this increase in the number of women law students feeds into greater numbers of women entering the legal profession. For example, the proportion of women trainees peaked in 1991–92 at 55 per cent, but since then has fallen to 53 per cent in 1996–97 at a time when the percentage of women law students has been increasing.[19]

Table 1.1: Applicants and acceptances for university first degree law courses in England and Wales 1994–96[20]

	Women applics.	Total applics.	Women accept.	Total accept.
For entry in:				
1994	11,548 (54 %)	21,266	4,883 (54 %)	8,997
1995	11,195 (56 %)	20,050	5,218 (56 %)	9,382
1996	10, 716 (58 %)	18, 595	5,769 (58 %)	9,998

Finding out about the representation of ethnic minority students, and, in particular, ethnic minority women students, is more difficult. For example, there is no information available about the comparative application and acceptance rates for ethnic minority students onto law degrees.[21] What is known is that of those who were accepted onto law degrees in 1996, one in five were from ethnic minorities, of whom just under two thirds were women,[22] representing a slight increase from 1995.[23] These figures compare favourably with those relating to the student body as a whole, which show that in 1996–97, just over ten per cent of full time undergraduates were from ethnic minorities.[24]

GENDER AT LAW SCHOOL

Learning to think like a (woman) lawyer

The core curriculum

Most law degrees cover what are known as the foundational, or 'core', subjects of law: criminal law, public law, tort law, contract law, European Union law, property law and equity and trusts. A student passing examinations in each of these areas will be awarded what is known as a 'qualifying law degree', meaning that she is exempt from the first part of the training required to become a solicitor or barrister. Appendix 1 outlines the different types of law degrees available and considers some of the reasons for choosing one option rather than another. For example, some universities will study law in a more technical scientific manner than other universities which may be keener to consider the law in its wider social, economic and political context. The former, more technical, approach to the law is also as the 'black-letter' law tradition.[25] This is a tradition which believes that the law is out there to be discovered: that there is an answer to every problem. Thus, it is often considered that the law school's function is to identify the legal principles that exist, and to inculcate them into students. Critics of this view of the law and law teaching argue that it is mostly devoid of considering the law in its social, political and economic context, and, therefore, in effect, the reality of the law.

Many writers have argued that the dominance of the 'black-letter' tradition is achieved by means of teachers 'convincing students that legal reasoning exists', that there is a 'legally correct' answer, that the law is objective and that emotion does not have a place in the law.[26] Thus, a 'failure to see the objectively right answer by the student is characterised by the lecturer (and frequently by the student) as a failure to learn the language of the law or a failure in knowing the correct rules or an inability to disentangle one's emotional or political sympathies from one's intellectual understanding of the law'.[27] This was borne out by the experiences of Caroline, a law student, who took a course called 'Feminist Perspectives in Law'. She says that she had 'expected more legal analysis, so I found myself tending towards a feeling of not doing what I "ought" to be doing in a law course'. When Caroline read law in the course title, she says that she expected 'black-letter law' which is the 'expected mould of an undergraduate taught law course'.[28] There is a movement away from this tradition, towards a greater recognition that the outcome of cases are determined not only by the content of the rules themselves but by the policy preferences of those who deal with them. This change reflects sustained criticism, not least from feminist scholars, and is exemplified by moves to consider increasingly the law from a gender/feminist perspective.

Integrating gender

The 'black-letter' tradition was criticised by Susan Atkins and Brenda Hoggett (now Mrs Justice Hale) in their ground-breaking book *Women and the Law*, in which they commented that their 'traditional legal education [had] been, in many ways, a hindrance to our understanding' of law's treatment of women.[29] 'Women', as a subject for, and a perspective from which, to study law began to enter the law school curriculum in the 1970s with the development of 'women and the law' courses. Such studies of 'women/feminism and the law' considered those areas of law which might be seen as having an immediate resonance with women's lives, covering issues such as the historical situation of women, feminist theories of law, pornography, rape, sexual violence and sex discrimination in the workplace.

Further developments in feminist legal thought have focused on questioning the entire way in which the law is constructed and developed, and law's claim to neutrality and objectivity is examined and its partiality revealed. This 'feminist jurisprudence' focuses not just on those substantive areas of law which might require reform, such as aspects of the law of rape, but considers the means by which laws are made, interpreted, enforced and discussed. Still later courses began to focus more on 'gender' rather than exclusively on 'women'. What unites all these courses is their attempt to draw together themes from across the substantive courses common in the law curriculum, to detach them from their supposed objectivity, to consider how the law impacts on women, and perhaps attempt to redress the way in which some of these issues are often mishandled in more mainstream or core courses.

These moves towards an understanding of the gendered nature of the law have been developing slowly in law schools. A survey of the law departments in 'old' universities in 1993 found that seven universities offered courses in Women and the Law.[30] A similar survey three years later, covering all universities, found that eleven universities offered a course 'Gender and the Law' and four 'Feminism and Law'.[31] What we do not know from these surveys is whether it is the same 'Women and the Law' courses which have metamorphosed into gender/feminism and the law.

Research carried out for this book, together with the surveys referred to above, would seem to suggest that there has been an increase in the number of courses offered, which is promising. For example, this new research found that 24 universities presently offer courses which consider the law from a gender/feminist perspective, the subject matter which ranges from 'Women and the Law', to 'Gender and the Law', to 'Feminism, Post-structuralism and the Law' (see appendix 2). A further two universities have run similar courses in the past, but have not done so for a few years. While the improvement in numbers augurs well for the future, there is, however, no room for complacency given that such courses are not available in around two thirds of law schools.

A potential hazard with such courses is that, being optional and not part of the 'core' of a law degree, they might continue to effect a marginalisation of gender insofar as certain areas of law are categorised as 'women's subjects' and not of concern to mainstream legal studies. It might be thought that the gender of law had been covered in the separate course and therefore did not need to pervade other options.[32] As Anne, a student of a feminist perspectives in law course, says, her one reservation about the course was that it placed the 'feminist perspective' of the law on the margins of legal study. Such issues should, she says, by rights, be compulsory for all law students.[33] The alternative to a specialised course is to integrate these issues, a view endorsed by the Australian Law Reform Commission (ALRC) which recommended that the 'experiences and perspectives of women should be included in all aspects of the law school curriculum'.[34] It also recommended that staff be given assistance in developing expertise in the field of feminist legal studies in order to be able to ensure that their teaching covers such perspectives.[35] Others, however, have argued that this debate sets up a false dichotomy, as a specialist course deals with issues and approaches which transcend a number of law courses and therefore would not otherwise be considered.[36]

New research for this book on the extent to which gender/feminist perspectives have been integrated into other courses found that just under half (45 per cent) of law schools said that they offered at least one course which included a gender/feminist perspective, the most common approach being to include feminist jurisprudence within the general jurisprudence course (see appendix three). This information is indicative only, in that no attempt was made to evaluate the role of the gender element to the course, and this brings to mind the cautionary note played by Carol Smart that feminist jurisprudence 'has not been taken seriously' in traditional jurisprudence courses.[37] Some progress, therefore, has been made, but much remains to be achieved to move close to the Australian position where 'most law schools now include feminist legal theory in the curriculum'.[38]

These issues need to be taken up at a national level, moving beyond individual struggles in institutions, perhaps drawing on the Australian experience. In 1993 considerable public concern over judicial attitudes to women was sparked off by the comment from South Australian Supreme Court Judge Justice Bollen that a 'measure of rougher than usual handling' may be used by a man to obtain his wife's 'consent' to sex.[39] The Australian government recognised the clear relationship between the cultural attitudes of the law and legal profession and the nature of law school curricula, and that the lack of any gender/feminist perspective in the curriculum was having an adverse effect on equality for women. Accordingly, it funded an inquiry into gender and the core law curriculum, resulting in the publication of a report and teaching materials.[40] The aim of the materials is to assist law students to think more laterally about legal problems, and to make women, and the multiplicity of women's concerns, more visible to law students and those teaching them.[41] These materials not only cover legal 'issues', but also attempt to broaden the roles of women within the law, and increase the visibility of women in legal debates and scholarship.[42]

While use of the Australian materials is not compulsory, the example set is important. In its 1996 report into legal education, ACLEC made no reference to the types of issues considered by the Australian authorities. The ACLEC report is promising in its emphasis on ethical values, recommending that law students 'must be made aware of the values that legal solutions carry, and of the ethical and humanitarian dimensions of law as an instrument which affects quality of life'.[43] In a sense, this is a recognition that the 'hidden curriculum' – the set of values, attitudes and the culture of the subject of study – is as important as the taught one. However, the report disappoints in failing to consider that a legal education which does not take into account the interests and perspectives of half the population, nor the ways in which the culture of the law and legal profession are gendered, can only consider the ethical and humanitarian dimensions of the law in a partial manner.[44] ACLEC's declaration that those involved in legal education have a responsibility to ensure that the law and legal profession reflects the social and cultural diversity of today's society, is at odds with its failure to argue the case for a more diverse legal education.[45]

A guide to equality in higher education published by the Equal Opportunities Commission (EOC), the Commission for Racial Equality (CRE) and the Committee of Vice-Chancellors and Principals (CVCP) is more advanced in its thinking, suggesting that, in order to ensure equality, universities must consider the content of curricula, and for example, 'how and to what extent has new scholarship on ... equal opportunities, gender, race/ethnicity been incorporated into course content' and whether 'resources and teaching materials are regularly checked for gender, racial and other forms of bias'.[46] The guide specifically refers to the teaching of law in the context of asking what preparation and support is available for staff dealing with potentially contentious 'equality issues' in the syllabus. This, of course, presumes that the law course has moved from first base and is actually contemplating the teaching of 'equality issues'.

While none of this is to suggest that legal education should be solely directed at the training of future lawyers,[47] even those students who do not pursue a career in the legal profession, become citizens with a view of the law, and on present terms it is likely to be a partial one. Equally, all law students are members of society and, as Martha Nussbaum has argued, they only become cultivated members of our community if their education takes into account issues relating to gender and women's experiences.[48]

In summary, progress, albeit slow, has been made in the UK. There is a great breadth of feminist legal scholarship which covers many areas of law and jurisprudence and the recently published *Feminist Perspectives on the Foundational Subjects of Law* represents a further step forward.[49] The focus must however be on materials for law teaching and this requires a thorough review of the law school curriculum, assisted by the type of institutional and professional support given in Australia.

Gender, language and stereotypes

Allied to the above discussion is the issue of gender stereotyping in law schools and the language used by students and staff in the context of class discussions and teaching and examination materials.[50] Language reflects social attitudes and mores and can have the effect of sustaining certain cultural attitudes: equally, changing and developing language has the capacity to help alter attitudes. For example, it is common sense, but has also been proven by detailed research, that the use of the male pronoun, or male suffix, such as chair*man*, conjures up the image of a man.[51] In educational terms, the repeated reference to a judge as male, whether in class discussions, teaching materials or exam questions, reinforces the notion of the judge as man, and men as the authoritative legal figures. Thus, the student who is continually confronted by assumptions that lawyers and judges are men, and perhaps that all litigants, clients, employers, employees and professionals are men, can begin to assume that it is men who rightly hold such positions and who are the (legitimate) users and authorities of the law; women being exceptional and not the norm.

Stereotyping may arise in general class discussions, teaching materials, and most often in what are often referred to as 'problem questions' or factual scenarios used to discuss legal issues in seminars, tutorials or exams. It is not difficult to find common gender stereotypes in teaching materials: very often there is a presumption of men as judges, lawyers and clients.[52] One extreme version of this was found in a sexual harassment scenario, the central characters of which were said to be Miss Nymph O'Maniac and Mr Gerald Grope.[53] The ALRC drew attention to the fact that 'sexist comments and stereotypes' used in class examples and examination questions have a potentially detrimental effect on women students.[54] It reported that in one contract law examination paper, there were fourteen roles, nine of whom were men, and four were women, with one unknown. Three out of the four women were ballerinas. The research also revealed how women were rarely in 'active' roles, but were usually secondary, and often attached in a relational way to the main (male) character, for example, as daughter or wife.[55]

The Canadian Bar Association, in its report on gender equality in the legal profession, echoes the ALRC's recommendation, in noting that all 'law schools should ensure that all aspects of ... legal education, including assessment tasks and course material, employ gender inclusive language and avoid existing stereotypes of the roles of women and men in society'.[56] Similarly, the UK bodies, the CRE, the EOC and the CVCP in their recently issued guide to equality in higher education recommend that teaching resources and materials be checked regularly for gender, racial or any other form of bias, and that guidance should be given to staff as to the use of inclusive language for all teaching purposes.[57] And, indeed, such recommendations simply replicate internationally agreed standards in the UN Convention on the Elimination of All Forms of Discrimination Against Women which requires state parties (which includes the UK) to eliminate 'any stereotyped concept of the roles of men and women at all levels and in all forms of education', involving in particular, the 'revision of textbooks' and the 'adaptation of teaching methods'.[58] Action should be taken immediately to implement these recommendations.

The Law Society's guide to the use of language for lawyers, *Clarity for Lawyers*, suggests that the use of gender-neutral language is essential to ensuring accuracy in writing, and that indeed it could be argued that in many ways language is *the* lawyer's tool. It also states that gender-neutral language is crucial to avoid discriminating against women, in that it is now taken as read that we should not make racist slurs, but that we still are prone to, at the least, 'belittling' women, and at worst being 'offensive'.[59] The guide lets itself down, however, by suggesting that the use of non-discriminatory language is 'not always easy to put into practice', although it is 'worth trying'.[60] We need a commitment to reform and change; to turn the rhetoric into reality. In this, we can learn from the US professional bodies which consider the use of discriminatory language, in certain circumstances, to be a disciplinary offence.[61] In addition, many Australian states demand gender-neutral wording in the drafting of laws and legislation,[62] and many states and provinces in the US and Canada have adopted express provisions on the use of gender-neutral language.[63]

Reform is essential to progress given the link between non-discriminatory language and the interpretation and implementation of the law. Feminist legal scholars have long pointed to the fact that behind the use of masculine language, the 'reasonable man' of tort law for example, lies a masculine interpretation of the law which may be inimical to the interests of many women.[64] Thus, the influence of language and stereotypes goes beyond having an educative function, to being crucial to the interpretation of the law itself.

Degrees of success

In 1995 and 1996 women achieved slightly more first and upper second class law degrees than men, and were less likely to be awarded a third class degree (see table 1.2).[65] However, the picture is not so clear and satisfactory regarding the proportions of women being awarded first class degrees. In the university sector as a whole, women are awarded far fewer firsts than men.[66] As regards women law students, in 1996, women were awarded 47 per cent of all firsts, representing a drop from 51 per cent in 1995.[67] Although the numbers are relatively small (341 firsts awarded in 1996, 160 to women), what is interesting is that the number of firsts awarded has increased, with proportionately more going to men.[68]

Further, figures from 1993 which relate to 'old' universities and 1991 ('old' and 'new' universities) show that in the former polytechnics, women were much more likely than men to be awarded firsts.[69] Thus, when the figures for 'old' and 'new' universities are collated, the fact that women have been awarded greater numbers of firsts in the polytechnic sector begins to mask the differentials in the 'old' universities. This might also be evidence of the trend apparent in respect of all students, whereby the largest discrepancy between women and men students, regarding first class degrees, is found at Oxbridge.[70] If it is the case that the number of firsts awarded to women is lower in the 'old' universities, the question has to be raised as to why this is the case. Is it due to the fact that, traditionally, the new universities have had a far greater commitment to equality which is reflected in both the performance and assessment of women law students?[71] Or, is it due to some other factor?

Table 1.2: Class of law degree awarded to women as a percentage of the number of degrees awarded in each class, 1991–1996[72]

Class of degree	1996 %women	1995 %women	1993 %women (univ.)	1991 %women	1991 %women (poly.)
First	47	51	44	48	58
Upper Second	56	56	53	52	61
Lower Second	52	51	49	48	50
Third/pass/other	42	40	46	52	47

It is good to report that the universities of Oxford and Cambridge are sufficiently concerned about the differentials in their degree awards that both institutions have funded an extensive research programme.[73] Preliminary findings raise more questions than answers, by suggesting that it is the 'natural caution' of women and their 'lack of confidence' that is preventing them from achieving firsts,[74] together with high anxiety levels among women shortly before exams which leads to a more cautious style.[75]

The EOC has rightly suggested that the reasons for differential achievement between women and men are unclear and further research is needed, but that the aim should be to improve the performance of both sexes to reduce the gender gap.[76] The EOC recommends that all universities collect data and set targets, suggesting that this form of 'gender monitoring' will be a source for policy making. Without this detail, it is difficult to determine what the problems are, or where or how to take action. The EOC also notes that much evidence has been produced in respect of primary and secondary education to show that teachers treat male and female pupils differently. Institutions need to be more aware that the personal bias of teaching staff can have an important effect on the manner in which they relate to students and in the way in which they present material. Universities should consider and address, through training and guidance, the following areas where teachers' own bias and prejudice can negatively influence the education they provide: teacher interaction with students, the delivery and content of courses, including subject-specific content and course culture, and assessment and examination of individuals.[77]

If change is to be achieved in university culture, the professionalisation of teacher training for university teachers is, through the work of bodies such as the Institute for Learning and Teaching in Higher Education, an excellent opportunity to enhance and improve training on gender and equality issues to ensure that the teachers of the future do not assume and repeat the prejudices of the past.

Women legal academics: role models and mentors

Fundamental to many of the issues discussed above is the fact that right across the university sector, women are under-represented at senior levels, and this is particularly true in law schools. Research carried out for this book reveals that women comprise 39 per cent of the total academic staff in UK law schools, although only 14 per cent of professors.[78] This is of fundamental importance to the culture of the law school. Students are generally taught by men, the senior academics are generally men, and the writers of scholarly articles (especially those generally employed in teaching) are therefore men. In an already masculine discipline like law, this simply serves to reinforce traditional views of women and of the law. It is likely to have a crucial impact on the way in which the law is taught and examined, on the content of the law school curriculum and on the experience of women law students. Indeed, the research from North America and Australia considered below suggests that the male domination of law schools adversely affects women law students.

The ACLEC report specified that one of the aims of legal education must be 'learning to act like a lawyer'.[79] If women students are to learn to act like lawyers, but the only role models they have are junior women, or women who are marginalised, they may themselves be imbued with a sense of the woman lawyer not as authoritative lawyer, judge or academic, but as handmaiden.[80] Or, where the dominant messages of what it is to be a lawyer are masculine, the emphasis may be on having to be masculine to be authoritative and successful. The absence, therefore, of women law teachers sends important messages about the nature of the law to law students. Accordingly, when William Wilson and Gillian Morris suggest that the object of legal education should be 'producing *tomorrow's* judges rather than *today's* trainee solicitors',[81] it must be an explicit aim that legal education makes it clear that tomorrow's judges can be women. And this cannot just be assumed. As Mona Harrington has observed, 'with virtually all-male faculty, the message that femininity and authority can be combined is not one that women law students frequently receive'.[82]

Such concerns apply equally to the understandings and expectations of male students. If male students are also gaining the sense of the lawyer as man, becoming and acting like a lawyer is going to be seen as copying the traditional stereotype of the lawyer, with conservative visions of work, women, the law and authority. Moreover, changes to the culture in, and attitudes which prevail at, law school and in legal practice are only likely to occur with the support and activism of both women and men.

Women law students: 'becoming gentlemen'?

US studies have suggested a number of ways in which the experiences of women law students differ adversely to those of their male peers.[83] First, the research suggested that there were strong attitudinal differences between women and men when they entered law school, with, for example, women citing a greater desire for public interest work, and a greater number of men expressing interest in financial gains and politics. It was found that these different aims homogenise by graduation to the extent that all students are more inclined to perceive their aims in accordance with those originally attributed to men.

A second finding was that in formal classroom instruction men were called on more often in class, volunteered more often, spoke for longer and received more attention from teachers. In addition, women were often subject to sexualised insults in the classroom. Allied to this was the fact that women law students tended to be more alienated from the informal educational networks of law school; for example, women were less likely to be mentored by academic staff. This, the researchers suggested, was partly due to the small numbers of women academics, with the implication that support and mentoring by the male academics tended to be directed at male students.

The final area, which is of most concern, was the finding from Pennsylvania Law School that, despite identical entry-level qualifications, by the end of their first year, men were three times more likely than women to be in the top 10 per cent of their law school class. This disparity continued to graduation so that women were graduating with 'significantly less distinguished professional credentials'.[84] The researchers put these differences down to the factors referred to above and in particular the general view that the teaching environment is a hostile one for many women. Accordingly, it was argued that there is no coincidence between being told that 'to be a good lawyer, behave like a gentleman'[85] and women feeling alienated by the law and law school.

We need to be cautious in our interpretation of the above studies. First, the law school experience of the US, as compared with the UK, is different in a number of ways. The US law degree is a postgraduate qualification, although it might be suggested that if women are coming to a degree at an older age, they ought to be less likely to suffer the alienation and diversion from originally expressed aims. Second, the teaching style said to give rise to alienation among women students, the Socratic method, involving class participation in lectures largely revolving around the discussion of cases, is not common in the UK. Nonetheless, the research was not exclusively based on Socratic style teaching. Third, the students who took part in the research graduated in or around the mid to late 1980s. In the last decade, women have begun to advance in the legal profession and academia, which might be expected to be reflected in any similar research carried out today.

Nonetheless, complacency would be misguided. The ALRC's inquiry into equality before the law details the alienation of women law students who perceive that they are 'not listened to in class or that their comments are not attributed the same weight as male students'.[86] This experience is compounded by the use of 'sexist comments and stereotypes in class examples and examination questions', as well as the fact that the 'experiences and perspectives of women are lacking in course materials and textbooks' (see above).[87] The report identified the small number of women academics, especially in senior positions, and the lack of credibility often given to women academics, as having 'serious implications for students in reinforcing the notion of women's inferior position in professional life'.[88] Margaret Thornton's work on Australian women law students also reveals similar feelings of alienation and concludes that, despite the fact that approximately 50 per cent of law students are women, 'they are not accepted at law school on equal terms with male law students, even though they perform at least as well educationally'.[89]

The very fact that the Australian research was carried out in the early 1990s, and relates to a similar legal education programme to that in the UK, provides, along with the US evidence, an agenda for action in the UK. The US research has principally been assisted by a variety of law schools, at the instigation of feminist scholars, supported in the context of nationwide inquiries into gender discrimination in the legal profession.[90] The Australian research was funded by government bodies. This issue needs to be given a considerably higher profile and priority in the UK. As it is central to any understanding of the role and status of women in the law, legal profession and judiciary.

COMMON PROFESSIONAL EXAMINATION

It is possible to become a solicitor or barrister having taken a non-law degree, or a law degree which is not a 'qualifying law degree', and the first step is the Common Professional Examination (CPE). The CPE is a one year full-time postgraduate course (two years part-time or distance learning) which completes the academic stage of the training necessary to become a practising lawyer, and consists of studying the seven foundational law subjects. Competition for places is generally intense, although it does appear that the number of those applying for the CPE has reduced in recent years.[91] Consequently, the 'applications to places' ratio for full-time places was 16:1 in 1993, falling to 8.1 in 1996.[92]

Women outperforming men?

As with degree results at undergraduate level, it would appear that women perform better in the CPE than men. Law Society research found that 62 per cent of students were awarded a pass, with 34 per cent gaining a commendation or merit, and less than two per cent a distinction.[93] Although the research found no significant variation between grades on the basis of age, ethnicity, school type, or institution, there has an overall difference between women and men. Women performed significantly better than men: 40 per cent of women gained a distinction, commendation or merit, compared with only 31 per cent of men.[94]

The nature of legal education and the CPE

There is an increasing interest and concern regarding the CPE route to the legal profession, fuelled by the fact that increasing numbers of training contracts and pupillages are being offered to CPE graduates rather than law graduates. As many as one third of students on the Legal Practice and Bar Vocational courses have completed the one-year CPE course.[95] Many City solicitors' firms offer training contracts to equal numbers of law and CPE graduates.[96] Furthermore, proportionately, CPE graduates have a higher success rate than law graduates in gaining a place on the Legal Practice Course (LPC).[97] This difference persisted even when academic performance had been taken into account, thus indicating that there was bias in favour of CPE students.[98] This bias in favour of CPE students exists also in the allocation of training contracts.[99] Although there was no evidence of bias in the award of places on the Bar Vocational Course (BVC), this bias was evident in the securing of a pupillage.[100]

It is not entirely clear why the legal profession appears to prefer a graduate who has studied law for one year (the CPE graduate) to the law graduate who has studied law for three years, sometimes four. There is much anecdotal evidence that somehow CPE graduates are 'better' lawyers.[101] But the question is 'better' in what sense. The fees for the CPE range from £2,500 to £3,500, in addition to the fees for the subsequent LPC or BVC, moving Peter Birks to ask, does 'better' mean 'better off, better connected [and] more congenial to the culture of the established practitioners?'[102]

These concerns appear to be substantiated by Law Society research which notes that this bias has serious implications for the social and ethnic composition of the legal profession. This is because those who had taken the CPE were, in general, drawn from 'more privileged social backgrounds than those who had done a law degree (they were more likely to have attended an independent school and had more highly qualified parents) and were more likely to be white'.[103]

One final concern with the growth in the number of CPE graduates is the nature of the legal education received. This chapter has already considered the nature and content of the law school curriculum, and the tendency for the 'black-letter' tradition to dominate, especially to the exclusion of any examination of the law from a gendered perspective. The CPE course, in its study of the seven foundational subjects of law only, and with an emphasis on the practical application of that law, as opposed to a critical appreciation of the law's context and impact, is the archetypal black-letter law course. Thus, the concerns raised in relation to the undergraduate law degree apply with even greater force to the CPE. Accordingly, as more and more CPE graduates enter the legal profession,[104] there is a possibility that this will not only adversely alter the composition of the profession, in terms of gender, ethnicity and class, but also that there will be little understanding of the role of law in perpetuating and reinforcing gender differences and disadvantages.[105]

POSTGRADUATE LEGAL STUDIES

Postgraduate legal studies offer a breadth of opportunities to continue studying law, perhaps motivated by a desire to study further a topic covered only briefly in a law degree, or a wish to study an area omitted from undergraduate studies. It can involve being taught, albeit at a more in-depth level than as an undergraduate, or can be self-directed through research; it can last for one year or up to six or more; it can be at the same institution as an undergraduate degree was taken, elsewhere in the UK, or anywhere in the world. Postgraduate studies can offer a sense of personal fulfilment and a gateway to the legal profession and academia.

Despite the clear advantages and opportunities which postgraduate legal studies offer, the processes by which such studies are undertaken are often shrouded in mystery. Although most universities provide careers guidance, postgraduate studies are often given little prominence. This section will seek to remove some of this mystery and consider some of the issues facing postgraduate women in law schools.

Choosing an institution and course

There are a variety of different postgraduate degrees in law, and full details of the different types of degrees are considered in appendix 4. In deciding to embark on a postgraduate degree, there are many factors to consider. There may be some advantages to choosing a different institution than where a first degree was taken, such as: exposure to different intellectual perspectives; experience different styles of teaching and assessment; mix with different people and make new friends; extend your range of extra-curricular involvement; and demonstrate flexibility and an ability to adapt to a new environment. Having said that, it may be that staying on at the same institution eases financial burdens, support networks may be stronger, and there may be a particular member of staff who is particularly suitable to supervise the research to be undertaken.

Particularly if considering a research degree, the choice of supervisor will be important, and may dictate the choice of institution. If no particular supervisor is known, other factors to take into account may be the research standing of the institution and the nature of the research that it carries out. For example, many departments specialise in certain fields of research, often demonstrated by the existence of a research unit/centre/institute. For example, the University of Bristol has a Centre for Law and Gender Studies which provides a focus for research in this area. Similarly, the University of Keele has a Gender, Sexuality and Law Research Group which provides a focus for research, conferences and seminars in this field. Appendix 3 lists all university law schools in the UK (with a few exceptions) and includes information as to whether individual law schools offer a 'gender and law' LLM, and/or an option in gender/feminism and the law at postgraduate level.

Making an application

Admission to postgraduate degrees is less structured than at undergraduate level and the methods of application will vary from institution to institution, and are not particularly rigorous or formalised. First contact may be made with the admissions officers of the various institutions who will provide prospectuses, application forms, funding details and any other information relating to the course or the applicant's particular needs. Applications are made separately to each institution, and an application for a research degree is more than likely to require a proposal setting out the area of research to be studied. Deadlines not only for admission to the particular degree, but crucially for scholarships and funding must be borne in mind, the latter often being set well in advance of the former.

As regards the factors which are considered to be important by universities, a survey in 1996 of university law schools found that 98 per cent of law schools said that the classification of the applicant's first degree was important or very important, with the academic reference being equally important. 80 per cent rated the research ideas of the applicant as important or very important; whereas 50 per cent similarly rated access to funding.[106] As with the factors considered relevant for undergraduate admissions, particular emphasis was given to 'motivation' and 'enthusiasm'.[107] The informal nature of admissions and the use of particularly subjective criteria raise similar problems to those identified regarding undergraduate admissions. Fair, open, accountable, monitored and evaluated policies are required, with appropriate training for all those involved in the selection process.

Women and postgraduate legal studies

In 1996, the total number of students taking postgraduate law courses was 4,942.[108] The vast majority of postgraduate students were in the 'old' universities, with only 14 per cent enrolled at the 'new' universities in England and Wales.[109] Although, as noted above, more women than men are undertaking undergraduate law degrees, and achieving, on average, better degree results than men, fewer women then men pursue postgraduate legal education. Women constitute 48 per cent of both full-time and part-time postgraduate law students,[110] compared with 54 and 46 per cent respectively in undergraduate law degrees.

Why are fewer women choosing postgraduate studies? Is it because women students are not used to seeing women academics, or seeing women academics in senior roles (to which the student may aspire), or because they perceive that the academic women are marginalised in the law school? Evelyn Ellis, professor of law at Birmingham University, finds from her experience that although there are many capable women law students, very few seem attracted to postgraduate studies and academic careers, and she suggests that this may in part be due to the fact that there are few female role models to emulate in universities. Or is it that women law students feel isolated in a masculine law school environment which does not instil in them a desire to pursue the law further? Or is it some other reason totally unconnected with gender?

In the light of her research that women often experience law school as alienating, Margaret Thornton has argued that it is the less than positive experience of law school which may explain the fewer numbers of women postgraduate law students.[111] Furthermore, her research revealed that many of these destabilising experiences of undergraduate law school persisted through postgraduate studies, with the effect that 'humiliation, rejection and marginalisation have resulted in a high attrition rate for women enrolled in higher degrees by research'.[112] The low numbers of women postgraduate students means that there is a reduced pool of women from which future academics are drawn. In addition, if there are few women postgraduate students, compounded with low numbers of academic women, law students lack many positive role models which may encourage them to consider further studies and an academic career. These points underline the importance of the need for further research which will illuminate our understanding of the experiences and expectations of women postgraduate law students.

CONCLUSIONS

The nature of the law school curriculum raises important considerations for both the experiences of women law students and the student body as a whole. The legal profession is drawn principally from those who study law at university, and thus the attitudes created, condoned or provoked at law school are likely to remain with students as they progress into the legal profession or academia. Accordingly, if indeed there appears to be a cultural resistance to women in the law and legal profession, which will be considered in the forthcoming chapters, it is essential that we focus first on the law school experience for any answers and potential reforms. In the words of the Australian Law Reform Commission, legal education can be 'identified as both the source of, and a solution to, the inequality women experience before the law'.[113] Law schools could indeed be 'crucibles of change in the legal profession'.[114]

It is in the law school that the values, ethics and principles of the law and legal profession are first introduced, developed and inculcated. And, it is the law students of today who will become the lawyers, academics and judges of tomorrow and who will, therefore, exert a considerable influence on both the role and status of women lawyers, and on the ways in which the law itself interacts with women's lives. What is required is no less than 'fundamental change in ... teaching practices, institutional practices, and social organisation'.[115]

PERSONAL TESTIMONIES OF WOMEN LAWYERS

Clare Maher, law student 1994–97

Clare Maher recently graduated with a law degree from Newcastle University and is thinking about becoming a solicitor. However, having gone straight from school to university, she wanted to take a break before going on to take the Legal Practice Course and is currently working for the Legal Aid Board. She describes below her experiences of law school and how she came to make her career choices ...

'I do not come from a legal background at all. My father is a painter and decorator in Manchester and my mother is a housewife and from an early age I was always encouraged to do my best at school. At age eleven, I secured an assisted place at St. Bede's College, an independent school, and, after doing well at GCSE and A Level, I went to Newcastle University to study law.

When I chose to study law, it wasn't specifically because I wanted to become a lawyer. I was unsure what I wanted to do but I had been told that a law degree would provide a sound base, whatever career I ultimately chose to pursue. During my A Levels I had the opportunity to shadow a judge in the Crown Court for a few days. I found it very interesting and I was totally in awe of how the barristers held the attention of the jury and coped with being shouted at by the judge. I thought that I would never have the confidence to do that!

However, after getting involved with mooting at university, I found that I quite enjoyed presenting a case to a judge and having to think on my feet. I got a real buzz when I won my first moot. Sarah, my mooting partner, and I reached the semi-finals of Newcastle Law School's Junior Mooting Competition in our first year and, in our second year, we went on to win the Senior Competition. The Senior Final was held in the Old County Court in Newcastle and was presided over by His Honour Mr Justice Sir Ian Kennedy, along with a local solicitor. It was a great experience to moot in front of a High Court judge.

Following on from this, Sarah and I were elected Master and Vice-Master of Moots of the Law School for 1996–1997. During this time we tried to encourage more students, both women and men, to get involved with mooting, and, because a high proportion of the participants in the competitions were female, we tried to get some women lawyers in to judge the finals of the competitions to provide some role models for the female students. Her Honour Judge Mary McMurray judged one of the finals and it was fascinating to hear her tell of what it was like when she was called to the Bar in the 1950s. It turned out that 14 out of the 16 students that reached the finals of the Junior and Senior Competitions were women. They were all very determined and committed to succeeding. I believe that you only get out of your time at university (or anywhere, for that matter) what you put into it. I personally got a lot out of being involved with and organising the mooting because I got to know a lot of people and I got a great feeling of satisfaction from seeing my work produce results.

You need to have so much more than just a law degree these days to get any job. In the careers' journals you often read of the advantages of travelling and doing different things, and I think this is very true. I spent the summer of 1996 working in a kitchen on a boys' camp in America and it was a fantastic experience. I met so many different people and had a great time travelling around America and staying in various places after the camp had finished. Obviously, not everyone is able to do this, but I think that any experience, whether it be work experience in a firm of solicitors or work in a non-legal environment, provides something extra to make you stand out from everyone else.

Through my involvement in the mooting competitions, I also got the opportunity to talk to members of the profession and ask them for advice. You can also get this from doing work experience with firms of solicitors or chambers. You will find that most lawyers love to talk about themselves and this can prove very useful. I remember a barrister telling me once that age is no disadvantage in the law and I think that this is very true. I graduated in July 1997 and, instead of going straight on to the vocational stage, which, for me at the time was the BVC and the search for a pupillage, I decided to have a bit of time out, get a job and a bit of money, then give my full attention to applying for pupillage.

I worked over the summer of 1996 as a researcher for this book and after investigating further both branches of the profession, I realised that there would be great opportunities for me as a solicitor. I like the idea of dealing with clients on a daily basis, but I also hope to continue with some advocacy and to obtain higher rights of audience. I believe that eventually the two professions will and should merge. Solicitors are starting to take on much more advocacy and there is no reason why they should not. I personally like the idea of taking a case from the beginning to the end and not having to pass it on to counsel half-way through, when I know that I could represent my client just as well as counsel could.

At the moment I am working at the Legal Aid Board doing the financial assessments on applications for Legal Aid and hopefully in the near future I will be doing the assessments on the legal merits of the applications. I plan to continue with this until the summer of 1999 and save up towards the tuition fees for the LPC. I am also trying to obtain a training contract before I commence the LPC. I know a few people who have done the LPC and cannot obtain a training contract, so are left with thousands of pounds of debts.

For my part, one of the main reasons for postponing the LPC was to have a real think about whether I wanted to make the commitment to it, because there is no way that my parents can help me with money. I do not want to work for a big city firm (quite often the only ones able to help with funding of the LPC). I think I am probably best suited to private practice and family work, more likely than not Legal Aid work. I hope that when I apply for training contracts, the fact that I will have experience of Legal Aid and that I have worked to pay for the LPC myself will work to my advantage, but who knows.

I think that the best advice that I could give to anybody at university who is thinking about entering the legal profession is to think carefully about what it is that you want to do and what you are suited to. There is a lot of work involved in getting a training contract or pupillage. It requires a lot of determination not to get disillusioned and give up. It does help if you can show that you have 'done something extra', whether that is sport, travel, mooting or some other kind of activity. It also helps to know something about the profession that you are hoping to enter and especially the particular branch that you have an interest in. You must realise what your qualities are and have confidence in them. Don't feel put down if you don't succeed straight away – just put it down to experience and try again. Realistically, very few people have everything fall into their laps and, when you finally get there, it will so much more satisfying if you have worked hard to achieve your success.'

Sadie King, PhD student

Sadie is in her final year as a PhD student and describes below her background and experiences and offers some thoughts for those contemplating a postgraduate degree...

'Coming from a one parent, working class family, there was no expectation on me to undertake higher education. However, the proximity of my primary and secondary schools to the local university brought me into contact with many parents who were university lecturers. Consequently, I intended to go to university from an early age and I never developed a 'them and us' mentality, which inhibits others with a similar background. Without this early contact with the academic world, it is less likely that I would have regarded attaining a university education as inevitable.

My twin sister and I were the first in our large, extended family to undertake higher education. My sister went to Newcastle Polytechnic to do a BSc in psychology; I went to Sheffield City Polytechnic to do an LLB in law. We both had reservations about attending a polytechnic, as opposed to a university, due to the prejudice against them (this prejudice still persists now that they are 'new' universities). However, our doubts proved to be ill-founded.

I commenced my PhD part-time at the University of East Anglia, at the same time as my sister commenced a Physiotherapy degree there (much to the confusion of university staff!) A year and a half later I received a scholarship from Newcastle Law School to study for my PhD full time. I am now in my final year at Newcastle University.

My interest in law came about through a growing awareness of the injustice prevalent in society and the power of law to attempt to correct such injustice. A love of courtroom dramas undoubtedly went some way to idealise the role of law in meting out such justice. However, it was Shakespeare who was, to a large degree, my inspiration:

We must not make a scarecrow of the Law,
Setting it up to fear the birds of prey,
And let it keep one shape till custom make it
Their perch and not their terror.

Measure for Measure Act 2, Scene 1

Shakespeare identified that the law must be treated as a 'living' and evolving aspect of the structure of civilised society. This observation, particularly because it recognises the importance of legal research, captured my imagination and has stayed with me ever since.

Financial obstacles

On successfully completing my LLB, I was offered a place at Chester College of Law, to train to be a solicitor. I decided to defer entry for a year in order to pay my debts. Half-way through the year, I was presented with a bill of £4,500 in respect of my fees. It was clear to me that I could not afford to do both solicitor training and legal research: I had to choose one or the other. I chose research. I spent the next year applying for numerous scholarships for research funding both in Britain and abroad. These all failed. On visiting the bank manager, hoping to obtain a loan, I was sent away to ask my (non-existent) wealthy parents to support me, which was immensely infuriating.

At this point, feeling my future plans were becoming more and more unattainable, I stumbled across an impressive scheme run by the University of East Anglia to encourage staff to pursue higher education. As I was a full-time administrative clerk for the university, I was entitled to study for a PhD, part-time, at the university, without paying fees. After six months, I commenced a new part-time administrative post at the university, which entitled me to have one quarter of the fees waived. Working part-time meant that money was scarce, but I had much more time to dedicate to my studies. After one and a half years part-time at the University of East Anglia, my financial difficulties were finally over, when I was awarded the scholarship to study at the University of Newcastle.

I believe that my determination to study part-time for six years to finish my PhD was a contributing factor in my obtaining a scholarship. I know two other postgraduate students who have had similar experiences: they had no luck obtaining funding until they had studied for a year, part-time and self-financing. As such, if financial difficulties appear to prevent you from undertaking postgraduate research, commencing your studies part-time is a very good option to consider.

The financial burden of postgraduate study has meant that I have not been able to do all the things that I would have liked. I had to make sure that I made the right choices; choices which I have never regretted. I did not commence postgraduate study because 'it seemed like a good idea at the time' or because 'I couldn't think of anything better to do'.

PhD Advice

Undertaking legal research can be very rewarding. However, it must to be fully appreciated before commencing studies, that studying for a PhD can be a difficult, lonely and/or isolating experience. It is important to know that you are suited to undertaking a research degree: completing an undergraduate dissertation or a research LLM are useful ways to establish if this is the case. Your choice of research subject is also important to your success. You can never take too long to decide on the subject matter. Remember, it will be your constant companion for the next three years, possibly longer. Choose wisely! It is also worth noting that it is usually possible to transfer to an institution after having commenced a PhD elsewhere. This may be necessary for a number of reasons and need not be problematic: as already noted, I transferred to the University of Newcastle from the University of East Anglia.

A good relationship with your PhD supervisor is very important to your success: an unhappy or unproductive relationship with their supervisors was the main reason why a number of friends either abandoned their PhD studies or transferred to other universities. If you are financially independent, it is worth looking at different universities in order to ensure that the person who would be your supervisor does, in fact, have a detailed knowledge of the area within which your research falls, and in order to find the faculty/ department which appears to offer the best support network for its research students.

An academic career?

University departments are now rated by their research excellence, and funding is allotted accordingly. As finance has been linked so closely to research output, the research potential of the 'lecturer-to-be' is her most valuable asset. Often, obtaining a PhD is insufficient to prove research potential. Law schools often expect to see that you have actively began publishing whilst a research student. As I wish to pursue an academic career, I have made this a priority: I had a book review published in the *Modern Law Review* in my second year of study and I am currently writing a second book review and an article for a reference journal. However, it is important to remember that your PhD is your priority. Obtaining teaching experience, whilst a postgraduate student, is also invaluable if you wish to pursue an academic career. Many law departments offer their postgraduate students the opportunity to undertake small group teaching of undergraduate students. Teaching not only provides worthwhile valuable training, it also provides a valuable source of income.'

NOTES

1   Quoted in Lorraine Dusky, *Still Unequal – The Shameful Truth about Women and Justice in America*, Crown Publishers: New York, 1996, p 16.
2   Quoted in Hedda Garza, *Barred from the Bar – A History of Women in the Legal Profession*, Franklin Watts: New York, 1996, p 52.
3   John Jenkins and Danielle Walker, *Annual Statistical Report 1993*, Law Society: London, 1993, p 75.
4   For further reading see, inter alia, Simon Lee and Marie Fox, *Learning Legal Skills*, Blackstone: London, 1994 and Wade Mansell et al, *A Critical Introduction to Law*, Cavendish: London, 1995.
5   See further *The Mature Student's Guide to Higher Education*, UCAS Handbook, *A Student's Guide to Entry to Law* (each available from UCAS), Phillip Kenny, *Studying Law* (4th ed), Butterworths, 1998; *Getting into Law* (2nd ed) 1996, MPW Trotman and Co Ltd (12 Hill Rise Richmond, Surrey, TW10 6UA, £6.95); *Students, Grants and Loans: A Brief Guide to Higher Education* 1997–98 (Department for Education and Employment, Publication Centre, PO Box 6927, London, E3 3NZ).
6   A survey of admissions to university law schools in 1996 discovered that the average A level offer from 'old' universities was 25.1 points, with 18.9 the average for 'new' universities: Phil Harris and Martin Jones, 'A Survey of Law Schools in the United Kingdom 1996', 31 (1997) *Law Teacher* 38–126, pp 60–61.
7   Where interviews are held, they should follow a pattern of structured questions, and those involved should be adequately trained in equal opportunities: see further Maria Tzannes, 'Strategies for the Selection of Students to Law Courses in the 21st Century: Issues and Options for Admissions Policy Makers', 29 (1995) *Law Teacher* 43–63.
8   Robert Lee, 'A Survey of Law School Admissions', 18 (1984) *Law Teacher* 165, pp 172–173.
9   Supra note 6, p 64.
10 ACLEC, *First Report on Legal Education and Training*, 1996, p 42.
11 See further, Maria Tzannes, supra note 7.
12 The ACLEC report recommends that universities review their equal opportunity policies regarding admissions, ensure that all staff engaged in

admissions are properly trained and that the policies are monitored: supra note 10, p 45. These recommendations are also supported by the joint publication from the CRE/EOC/CVCP, *Higher Education and Equality: A Guide*, 1997, p 33 (available from the EOC, £5).

13   John Wilson, 'A third survey of university legal education in the United Kingdom', 13 (1993) *Legal Studies* 143–182, p 152.

14   Supra note 5.

15   Under the heading 'Gender at Law School'.

16   This is considered further in the following chapter under the heading 'A Woman's Place'.

17   See also chapter two, under the heading 'A Woman's Place', and the details of the methodology considered at the end of appendix three.

18   Under the heading 'Women legal academics: role models and mentors'.

19   Bill Cole, *Trends in the Solicitors' Profession – Annual Statistical Report 1997*, Law Society: London, 1997, p 71.

20   Verity Lewis, *Trends in the Solicitors' Profession – Annual Statistical Report 1996*, Law Society: London, 1996, p 67 and ibid p 61.

21   Although UCAS statistics provide information on ethnicity, including a breakdown of different ethnicities and by sex, these figures are not broken down in terms of the subject matter of law. UCAS: *Annual Report 1995 Entry*, UCAS: Cheltenham, 1996, pp 166–171.

22   Supra note 19, p 62.

23   Supra note 20, p 61.

24   HESA Press Release 14, 9 May 1997.

25   See further Fiona Cownie and Anthony Bradney, *English Legal System in Context*, London: Butterworths, 1996, pp 126–130 and the references cited therein.

26   Duncan Kennedy, 'Legal Education as Training for Hierarchy', in David Kairys (ed), *The Politics of Law*, Pantheon Books: New York, 1990, pp 43–44.

27   Supra note 25, p 128.

28   Quoted in Susie Gibson, 'Define and Empower: Women Students Consider Feminist Learning', 1 (1990) *Law and Critique* 47–60, p 48.

29   Susan Atkins and Brenda Hoggett, *Women and the Law*, Blackwell: Oxford, 1984.

30   Supra note 13, p 169.

31   Supra note 6, pp 51-52.

32   See further Mary Jane Mossman, 'Otherness and the Law School: A Comment on Teaching Gender Equality', 1 (1985) *Canadian Journal of Women and the Law* 213, p 214.

33   Supra note 28, p 58.

34   ALRC, *Equality before the Law: Women's Equality*, Report No 69, 1994, para 8.23. See http://www.anu.edu.au/alrc/.

35   Ibid para 8.25.

36   Regina Graycar and Jenny Morgan's book, *The Hidden Gender of Law*, Federation Press: Australia, 1990, seeks to break away from the traditional categories of law and examine the law 'taking women's lives as the starting point' (p 3) an approach which is not likely to be taken in mainstream courses.

37   Carol Smart, 'Feminist Jurisprudence', in Peter Fitzpatrick (ed), *Dangerous Supplements*, Pluto Press: London, 1991, p 133. Although others have since

been more positive, suggesting that feminist jurisprudence has taken a 'secure hold' within law schools: Hilaire Barnett, 'The province of jurisprudence determined – again', 15 (1995) *Legal Studies* 88–127, p 120.

38 Supra note 34, para 8.20.

39 Quoted in Regina Graycar and Jenny Morgan, 'Legal Categories, Women's Lives and the Law Curriculum', 18 (1996) *Sydney Law Review* 431–450, p 431.

40 To be found at http://uniserve.edu.au/law and discussed ibid.

41 Supra note 39, p 437.

42 Seeking to reverse a trend identified, for example, by Mary Joe Frug's analysis of a contract law casebook which noted that women were all but absent from the book, except where they were portrayed in stereotypically weak roles: Mary Joe Frug, 'Re-reading Contracts: A Feminist Analysis of a Contracts Casebook', 34 (1985) *American University Law Review* 1065.

43 Supra note 10, p 17.

44 One further value which legal education is, according to the ACLEC report, to develop is 'equality of opportunity'. Questions, however, must be asked as to how this is to be achieved when there remains an absence of academic women in law schools, particularly at senior levels, together with a tendency to assume that the actors in the law and legal profession are men and the exclusion from the curriculum of any wider consideration of gender or women is common.

45 Supra note 10, p 17.

46 Supra note 12.

47 The undergraduate law degree should be about education rather than training: see Dawn Oliver, 'Teaching and Learning Law: Pressures on the Liberal Law Degree', in Peter Birks (ed), *Reviewing Legal Education*, Oxford University Press: Oxford, 1994, pp 77–86.

48 Martha Nussbaum, *Cultivating Humanity: A Classical Defense of Reform in Liberal Education*, Harvard University Press: Harvard, 1997.

49 Anne Bottomley (ed), *Feminist Perspectives on the Foundational Subjects of Law*, Cavendish: London, 1996. The book contains two essays relevant to each of the seven foundational subjects of the law curriculum. Further edited collections are in progress which focus on specific areas of the law; with Health Care Law, Labour Law and Public Law collections being published in 1999. The texts are, however, academically advanced and although aimed at law students, may only be accessible to a limited number of students.

50 See further Mary Jane Mossman, 'The Use of Non-Discriminatory Language in the Law', 73 (1994) *Canadian Bar Review* 347–371.

51 Ibid p 350.

52 These points may apply equally to law school moots (mock trials), see for example Mairi Morrison, 'May it Please the Court?: How moot court perpetuates gender bias in the "real world" of practice', 6 (1995) *UCLA Women's Law Journal* 49–84.

53 Quoted in Hilary Sommerlad and Peter Sanderson, *Gender, Choice and Commitment: a study of women lawyers*, Dartmouth-Ashgate: Aldershot, forthcoming.

54 Supra note 34, para 8.28.

55 Ibid.

56 Canadian Bar Association, *Touchstones for Chance: Equality, Diversity and Change*, 1993, recommendation 8.1, p 30.

57   Supra note 12, pp 39–40.
58   Art 10(c). See United Nations Blue Books Series Vol VI, *The Advancement of Women 1945–1995*, UN: New York, 1995.
59   Mark Adler, *Clarity for Lawyers: The use of plain English in legal writing*, Law Society: London, 1990, pp 71–73.
60   Ibid p 73.
61   Hence, when a male lawyer referred to his opposing counsel as variously 'little lady', 'little mouse' and 'young girl', it was more than appropriate that he was fined by the relevant professional body for his inappropriate and unprofessional comments: supra note 50, p 352.
62   See Michele Asprey, *Plain Language for Lawyers*, Federation Press: Australia, 1991, pp 111–115.
63   Supra note 50, p 352.
64   See for example, Robyn Martin, 'A Feminist View of the Reasonable Man', 25 (1994) *Anglo-American Law Review* 334–374 and Regina Graycar and Jenny Morgan, supra note 32.
65   In both years, 51 per cent of women gained first or upper second class degrees compared with 44 per cent in 1995 and 46 per cent in 1996 for men: supra note 19, p 64 and note 20, p 69.
66   In 1995–96 while eight per cent of male students received a first class degree, only six per cent of women did so. In 1993-4, the figures were nine per cent and seven per cent respectively: Harriet Swain, 'First-class degrees on wane', *Times Higher Educational Supplement*, 4 July 1997.
67   Supra note 13, p 64 and note 14, p 69.
68   The number of men gaining firsts increased in 1996 by 31 per cent, the number of women increasing by 13 per cent: supra note 19, p 64 and note 20, p 69.
69   Supra note 3, p 78.
70   At both Oxford and Cambridge more than 16 per cent of men are awarded first class degrees, compared with only around nine per cent of women: Donald MacLeod, 'Confidence Trick', *The Guardian*, 13 January 1998.
71   An interesting research study would involve comparison of entry qualifications with degree results.
72   Supra note 3, p 78, note 19, p 64, note 20, p 69, and Law Society, *Trends in the Solicitors' Profession – Annual Statistical Report 1994*, Law Society: London, 1994, p 77.
73   Donald MacLeod, supra note 70; John O'Leary and David Charter, 'Oxford prepares the way for more women firsts', *The Times*, 29 January 1998.
74   John O'Leary, 'University girls fall behind in confidence', *The Times*, 5 January 1998. Perhaps being infantilised by being referred to as 'girls' also adds to the problem.
75   Supra note 70.
76   EOC, 'EOC Response to the National Committee of Inquiry into Higher Education', November 1996.
77   The joint EOC, CRE and CVCP guide to equality in higher education, supra note 12, recommends that 'resources and teaching materials are regularly checked for gender, racial and other forms of bias'.
78   See chapter two under the heading 'A Woman's Place', tables 2.1, 2.2 and appendix three.

79 Supra note 10, p 24–25.
80 Emma Coleman Jordan has argued that the role model argument ignores the 'dynamic, intergenerational nature of both teaching and learning' and thus the argument 'devalues the contribution that our students can make' and that it is 'mentors', who are continuously legitimated and reinforced by an interactive, communicative process that are required: 'Images of Black Women in the Legal Academy: An Introduction', 6 (1990-91) *Berkeley Women's Law Journal* 1–21, pp 7-8. I would suggest that role models and mentors are necessary. The mentor is the academic woman with whom the student has personal contact, be they tutor, teacher or other member of the law school. In that sense, the mentor is someone engaging not just in an exercise of 'look at me', but in a positive dialogue with students. On the other hand, a role model may be someone whose writings a student comes across or to whom reference in teaching and writing is made, or someone in the law school with whom the student has no contact. In other words, someone who is not known personally to the student and who cannot therefore engage personally. Role models and mentors are important for women students.
81 William Wilson and Gillian Morris, 'The Future of the Academic Law Degree', in Peter Birks (ed), *Reviewing Legal Education*, Oxford University Press: Oxford, 1994, p 104, emphasis in original.
82 Mona Harrington, *Women Lawyers – Rewriting the Rules*, Plume-Penguin: London, 1993, p 45.
83 For example: Lani Guinier et al, 'Becoming Gentlemen: Women's Experiences at One Ivy League Law School', 143 (1994) *University of Pennsylvania Law Review* 1–110; Janet Taber et al, 'Gender, Legal Education and the Legal Profession: An Empirical Study of Stanford Law Students and Graduates', 40 (1988) *Stanford Law Review* 1209–1297; Lee Teitelbaum et al, 'Gender, Legal Education and Legal Careers', 41 (1991) *Journal of Legal Education* 443–481; Catherine Weiss and Louise Melling, 'The Legal Education of Twenty Women', 40 (1988) *Stanford Law Review* 1299–1369.
84 Lani Guinier ibid p 21. The survey of the University of New Mexico law School, Lee Teitelbaum ibid, found no difference in academic performance between women and men. The difference between the two studies however is that the Pennsylvania study used statistics for all students across all years, rather than the reliance on self-reporting in the New Mexico study.
85 Quoted in Lani Guinier, supra note 83, p 5.
86 Supra note 34, para. 8.7.
87 Ibid.
88 Ibid.
89 Margaret Thornton, *Dissonance and Distrust: Women in the Legal Profession*, Oxford University Press: Oxford, 1996, p 104.
90 In addition, in early 1998 the American Bar Association (the regulatory body of the legal profession and the organisation which accredits law schools) published a document urging law schools to undertake studies to 'determine the level of gender-parity of students' experiences at their own law schools': see ABA Issues Guide to Make Sure Law Schools are Female-Friendly, http://www.abanet.org/media/jan98/graperel.html
91 Applications are made via a standard application form obtained from the CPE/ Diploma in Law Central Applications Board, the address of which can be found

in appendix five. Applications for part-time or distance learning courses must be made directly to the institution offering the course.

92   Phil Harris and Steve Bellerby, *A Survey of Law Teaching,* Sweet & Maxwell and the Association of Law Teachers: London, 1993, p 42 and supra note 6, p 71.

93   Michael Shiner and Tim Newburn, *Entry into the Legal Professions – The Law Student Cohort Study Year 3,* Law Society: London, 1995, p 27.

94   Ibid.

95   Supra note 10, p 69.

96   Peter Birks, 'Rough justice for law students', *The Times,* 14 September 1993.

97   Supra note 93, p x and Michael Shiner, *Entry into the Legal Professions – the Law Student Cohort Study Year 4,* Law Society: London, 1997, p 25.

98   Supra note 89, p x.

99   Supra note 93, p 53.

100   Ibid p 124.

101   Supra note 96.

102   Ibid.

103   Supra note 93, p xii and note 97, p x.

104   The ACLEC report recommended that conversion courses remain available and rejected calls for a two-year compulsory course on the basis that this would reduce access to the profession: supra note 10, pp 69–71.

105   Nonetheless, the CPE does makes available a legal qualification to many people who might not otherwise be able to afford a three year degree. For example, fellows of the Institute of Legal Executives may take the CPE (part-time) in order to begin the process of qualifying as a solicitor.

106   Supra note 6, pp 72–73.

107   Ibid.

108   Ibid pp 77–79. This figure covers courses in England, Wales, Scotland and Northern Ireland, and excludes the provision of CPE courses and diploma courses.

109   Ibid.

110   HESA Data Report, July 1995.

111   Supra note 89, p 104.

112   Ibid.

113   Supra note 34, para. 8.1.

114   Rand Jack and Dana Crowley Jack, *Moral Vision and Professional Decisions: The Changing Values of Women and Men Lawyers,* Cambridge University Press: New York, 1989, p 166.

115   Lani Guinier, supra note 83, p 100.

Chapter 2
Academic women in the law school

'Men have had every advantage over us in telling their own story. Education has been theirs in so much a higher degree; the pen has been in their hands.'

Anne Elliot in Jane Austen's *Persuasion*

'Institutions dedicated to the unraveling of truth are themselves still wrapped in the myths of the past.'

Report of the Hansard Society Commission on Women at the Top, 1990[1]

John Stuart Mill wrote in 1869 that it would only be when women entered the academy, undertook their own research and told their own story, that an understanding of women which was other than partial would occur.[2] Over one hundred years later, women have entered the legal academy, but we still await a full account of our experiences.[3] This chapter begins to remedy this situation. It starts with an analysis of the representation of women in the academy generally, and law schools in particular, followed by a discussion of some of the reasons why women remain marginalised and under-represented in universities, especially at senior levels. Recent changes and developments in the academic world are then examined to assess their impact on the representation and status of academic women in the law school. The chapter concludes with the personal testimonies of a number of women legal academics which exemplify and expand on many of the themes identified in this chapter.

A WOMAN'S PLACE

In the university sector as a whole, women account for only eight per cent of professors,[4] and 35 per cent of all other academic staff.[5] Research in 1998 by the Association of University Teachers (AUT) shows that out of the 178 higher education institutions, 20 had no women professors[5] and eight had no women senior lecturers or senior researchers.[6] As regards ethnic minority representation, only 0.6 per cent of women professors are from ethnic minorities (1.6 per cent of men), and these comprise five per cent of other academic staff (six per cent of men).[7] In the management ranks of universities, there are only six women vice-chancellors.[8]

But, not only are there few women, especially at senior levels, there is also stark evidence of inequality of pay between women and men academics. An AUT study in 1992 concluded that 'women are systematically paid less than men in universities'.[9] The report revealed that women were paid, on average, £3,249 less than men: or, 84 per cent of men's salaries. These findings were confirmed in a later study by Robert McNabb and Victoria Wass which concluded that 'even after controlling for rank, age, tenure and faculty, a gender effect in the remuneration of British academics remains'.[10] The average differential between women and men's salaries was found to be 15 per cent. Analysis of a cohort employed in 1975 and still working in 1992 revealed that more men than women were promoted, and that unless great differences can be solely accounted for in terms of differential productivity, a gender element exists in the promotions exercises. This study reveals not only the differences in the pay of women and men academics, but also an increase in the differential. This is a particularly worrying trend as it coincides with the progressive deregulation of pay and promotions structures in universities.

Unfortunately, comparative information on the representation and pay of women *legal* academics is not available. The first major survey of law schools and legal education, published in 1966, noted that there was an 'almost total lack of information concerning ... law schools'.[11] However, despite the survey, and a number of others updating it,[12] it continues to be the case that although we might now have *some* information, there remains a dearth of data concerning *women* in law schools. Although the omission of gender in the 1966 survey may be understandable, by the 1990s it is no longer.[13] Furthermore, the government body responsible for the collation of higher education statistics, the Higher Education Statistics Agency (HESA), does not collect information relating specifically to law schools.[14]

In an attempt to improve matters, a survey was undertaken for this book with the aim of collecting data on the representation of academic women in UK university law schools. While the survey represents no more than the first, albeit important, step towards improving matters, its findings provide a starting point for debate within each law school and across the sector as a whole, and a base from which to launch further quantitative and qualitative research.[15] In an age of 'freedom of information', universities, as publicly accountable bodies, must be open to analysis of their activities. Furthermore, as argued in chapter one, law schools are important institutions in terms of the development of the law and legal profession and therefore their activities must be open to scrutiny. The law schools must be praised for contributing so openly to the survey, thus helping to begin a process of assessment and a culture of self-evaluation which bodes well for the future of law schools. The exceptional response rate, of 93 per cent, for a postal survey of this kind, suggests that law schools are ready to embrace self-evaluation. The data displayed in tables 2.1 and 2.2 has been compiled from the information given by the law schools, without any alterations.[16]

Table 2.1 Number of academic staff in surveyed university law schools as at 1 October 1997[17]

	Men	Women	Total	% women
Professors	258	43	301	14
Senior staff	585	394	979	40
Lecturers	489	409	898	46
Total staff	1,332	846	2,178	39

The survey found that 14 per cent of law professors are women (see table 2.1). This is a positive result in that across the university sector as a whole, women comprise only nine per cent of professors. Nonetheless, there can be no complacency because, of the seventy-five law schools responding to the survey, three out of every five (63 per cent) had no women law professors. In those law schools where women are represented in the professoriate, only two law schools had more than two women professors.[18] The number of women legal academics improves lower down the hierarchy: 40 per cent of senior staff, which includes senior and principal lecturers and readers, and 46 per cent of lecturers are women. It augurs well for the future that over one third of staff in law schools are women.

The data provides some stark results, and care must be taken in assessing and interpreting the policies and practices in each law school. For example, a law school's station in table 2.2 may be affected by its size, history, starting point, or the fact that it may have recently implemented an equal opportunities programme which is only just beginning to bring results. Further, there may have been changes in the composition of a law school, since the survey's census date of October 1997, which would place it quite differently in table 2.2. Thus, readers are encouraged to explore the results with individual law schools in order to place the data in context.

What remains undiscovered is the representation of ethnic minority women legal academics. Neither the surveys of law schools referred to above, nor the survey carried out for this book, detail the representation of ethnic minorities on the academic staff of law schools. This is a regrettable omission in view of the low number of ethnic minority staff in academia as a whole. In 1990–91 a US journal published a symposium of articles on black

Table 2.2 Number of men and women academic staff in UK university law schools in order of percentage representation of women as at 1 October 1997

University	Men	Women	Total	%Women
Heriot Watt	1	3	4	75
Oxford Brookes	5	9	14	64
Middlesex	8	13	21	62
Hertfordshire	13	20	33	61
Thames Valley	14	22	36	61
Coventry	8	12	20	60
Sussex	7	10	17	59
Leeds Metropolitan	12	15	27	56
Lincolnshire & Humberside	4	5	9	56
Kingston	9	11	20	55
Surrey	5	6	11	55
Central Lancashire	16	19	35	54
Liverpool	12	14	26	54
Glamorgan	16	18	34	53
Glasgow	16	18	34	53
Plymouth	8	9	17	53
East London	16	16	32	50
Huddersfield	10	10	20	50
Robert Gordon	7	7	14	50
Sheffield	23	22	45	49

Bristol	27	24	51	47
De Montfort	23	20	43	47
North London	9	8	17	47
Sheffield Hallam	10	9	19	47
Wolverhampton	22	19	41	46
Liverpool John Moores	12	10	22	45
Keele	9	7	16	44
Westminster	18	14	32	44
Brunel	13	10	23	43
Manchester Metropolitan	31	23	54	43
Aberystwyth	17	13	30	43
Exeter	15	11	26	42
Napier	11	8	19	42
Nottingham Trent	29	21	50	42
Bournemouth	10	7	17	41
Central England	16	11	27	41
London Guildhall	37	26	63	41
Manchester	19	13	32	41
Derby	9	6	15	40
South Bank	12	8	20	40
Staffordshire	21	14	35	40
Durham	11	7	18	39
Kent	17	11	28	39
West of England	37	24	61	39

University	Men	Women	Total	%Women
Teeside	8	5	13	38
Buckingham	12	7	19	37
King's College London	29	17	46	37
Queen's	22	13	35	37
Dundee	15	8	23	35
Glasgow Caledonian	11	6	17	35
Reading	11	6	17	35
Leeds	19	10	29	34
Abertay	8	4	12	33
Lancaster	12	6	18	33
Northumbria at Newcastle	56	26	82	32
Southampton	19	9	28	32
Ulster	11	5	16	31
Warwick	24	11	35	31
Hull	19	8	27	30
Newcastle upon Tyne	14	6	20	30
London School of Economics	24	10	34	29
Nottingham	23	9	32	28
Essex	19	7	26	27
City	23	8	31	26
Birkbeck	9	3	12	25
Edinburgh	33	11	44	25

SOAS	15	5	20	25
Birmingham	28	9	37	24
Leicester	31	10	41	24
Strathclyde	19	6	25	24
University College London	31	10	41	24
Greenwich	21	6	27	22
Cambridge	47	12	59	20
Aberdeen	21	5	26	19
Cardiff	53	5	58	9
Anglia Polytechnic	-	-	-	-
Aston	-	-	-	-
East Anglia	-	-	-	-
Luton	-	-	-	-
Oxford	-	-	-	-
Queen Mary & Westfield College	-	-	-	-
TOTAL	**1,331**	**846**	**2,177**	**39**

women law professors, noting that: 'The story of black women law professors in the legal academy has yet to be told.'[19] If this is the case in the US, it is even more so in the UK.

UNDERSTANDING ACADEMICS

Roles for women

Australian law professor Margaret Thornton has suggested that although women have been allowed entry into the legal academy, this acceptance is equivocal.[20] Indeed, this acceptance, she argues, is granted only if women conform to certain roles or stereotypes. One of these roles is the 'body beautiful'; the woman academic who is expected to look sexy and satisfy the ego of 'benchmark' man. Not only does she flatter benchmark man, but the contrast between the two of them – one deferential and sexual, the other authoritative and rational – emphasises who is the serious and esteemed academic. Another role is that of the 'adoring acolyte' who strokes benchmark man's intellectual ego, praising him, preferably in her written work. Of course, the combination of 'adoring acolyte' and 'body beautiful' is particularly sought after. The 'dutiful daughter' serves the academic community well. She is the good 'campus citizen', with her large amounts of teaching, preferably on compulsory courses, her pastoral care of students, her useful organisation of conferences which assist the research careers of the real (men) academics, as well her membership of a plethora of relatively unimportant committees in order to give the appearance of gender equality. And, finally, the 'queen bee' is honorary man; deriding feminist women and scholarship, adopting the culture and attributes of benchmark man, again giving credence to the law school's philosophy of equality.

Although these stereotypes do not reflect the diversity of women in the academy, there is more significance to them than the fact that they engender a panic in the woman legal academic as to whether or not she falls within one of these roles, and then whether or not this is an advantage or disadvantage. What is crucial about the typology is that none of the roles is empowering, each emphasises what is feminine, ie appearance, sexuality, deference, diligence and self-sacrifice, and all of them ensure that women are kept constrained within the dominant (masculine) culture; marginalised, and of no threat to the hegemony.[21] In addition, it emphasises that increasing numbers of academic women in law schools only represents a start: whilst necessary, numbers alone are not sufficient to change a culture.[22]

It is the dominance of masculine cultures and attitudes which lie behind Margaret Thornton's argument. She suggests that academic women disrupt the sexual regime which is premised on men as authoritative knowers of the law, women as handmaidens.[23] Thus, to gain any acceptance, women have to try to conform to an existing male-dominated regime, disrupting the power hierarchies as little as possible. Furthermore, what is valuable about Thornton's work is that she articulates the ways in which the particular discipline of law may influence the professional life of academic women. The dominance of the 'black-letter' law culture, examined in chapter one, with its emphasis on the authoritative, unitary, objective and impartial nature of law, suggests such a degree of congruence with the image of masculinity that women as authoritative legal scholars is almost a contradiction in terms. This is especially so for the feminist scholar. As Thornton writes, '[i]f legal scholars have devoted their academic lives to the formalistic study of legal doctrine, they are likely to be nonplussed to be told that they are both perpetuating and constructing legal knowledge which is partial'.[24] Thus, although there are common features between women's experiences of academia in general and law schools in particular, there are also particular ways in which the subject of the law exerts a disciplinary influence on women legal academics.

Although drawing on the experience of Australian women legal academics, Margaret Thornton's work has many resonances with the UK and provides a useful, theoretical basis for understanding the experiences of academic women in UK law schools. This reasoning may help to explain why there are so few women in senior positions in law schools, and why, despite the existence of many women legal academics, this does not translate into equal status or recognition.

The equivocal acceptance of women academics may also be demonstrated by reference to the research on women's inequality of pay. The AUT report considered above concluded that women's lower salaries may be the result of women being appointed lower on the universities' salary scales than men of the same age, and that the difference between the pay of women and men is a reflection of the overall lower status accorded to women within academia.[25] The study by McNabb and Wass confirmed the AUT's analysis that women face lower returns for age, suggesting that either women academics are appointed at lower points on salary scales, or that they benefit less from accelerated progression through increments (or both).[26] Crucially, this study found differences between the disciplines, to the extent that where there were higher numbers of women, salary levels were less for both women and men. Thus, women employed in more male-dominated faculties do better than women in less male-dominated faculties.[27] This is strongly suggestive of women not having the same 'purchase' or 'cultural capital' as men, with the effect that the more women there are in a faculty or department, the less it is respected, with lower salaries for all.

Whether or not women legal academics are paid less than their male colleagues is not known. The only research which begins to touch on this issue was carried out by Anglia Polytechnic University in 1995.[28] The study found that flexible working practices, good quality administrative support, improved financial incentives, recognition by senior staff (head of department or above) and improvements in library resources were all of greater priority to women than men.[29] The authors therefore concluded that: 'Clearly, those most keenly looking to overall improvements in their work conditions and opportunities are women.'[30] Although this does not demonstrate inequality of pay, it does suggest that research into the working conditions of women legal academics may reveal a difference in expectations and rewards to those of men. In this context, the recognition by the Committee of Vice-Chancellors and Principals (CVCP) that there exists a link between the marginalisation of women and lower rates of pay is important. The CVCP has stated that because pay levels are determined by subjective standards which often reinforce sexual stereotypes, this may lead to lower ratings of women even though performance may be the same.[31]

This may also impact on promotions exercises. The law professor is traditionally associated with certain characteristics – knowledge, judgment, authority and masculinity – characteristics which are not generally associated with women. As Margaret Thornton argued above, there is much resistance to seeing women as authoritative purveyors of legal knowledge, resistance which may manifest itself in certain assumptions. For example, what is deemed to constitute 'quality research' (feminist research of any kind raising doubts), 'quality' teaching (an authoritative (male) figure perhaps being seen as 'better'), and who would best fit into the atmosphere and ethos of the department, who would least disrupt the present (sexual?) regime. Would this lead to the preference of a 'mediocre man over a bright, assertive woman'?[32] The Equal Opportunities Commission (EOC) suggests that 'the nature of the academic environment is one of a male hierarchy ... [which] will inevitably act as a barrier to the promotion and retention of women'.[33] In particular, the EOC criticises the secrecy and unprofessional manner of many appointments to senior posts in higher education. In a similar vein, the AUT has criticised the 'combination of secrecy, subjectivity and amateurism' which surrounds many promotions procedures.[34]

The Hansard Society's report *Women at the Top* emphasised that what was lacking in universities was not the numbers of suitable women waiting for preferment, but 'the persistence of outdated attitudes about women's roles and career aspirations' which 'constitute the main barrier stopping women from reaching the top in academic life'.[35] The report was damning in its description of universities as 'bastions of male power and privilege', and noted the deep irony that 'institutions dedicated to the unraveling of truth are themselves still wrapped in the myths of the past'.[36] One such myth of the past, which Baroness Helena Kennedy QC has suggested remains, is the belief that women are simply not of the same academic calibre as their male colleagues.[37] She continues that this fiction passes as a test of 'merit', but what constitutes merit reflects 'male standards and criteria'. This argument often has a biological element, in that women are seen to be at the mercy of their hormones, are often irrational, emotional, and therefore, clearly, not the objective, neutral and rational purveyors of legal knowledge that many men are. Another variant is that those who do excel are often seen as the exceptions which prove the rule.[38]

A further aspect to the dominance of certain stereotypes and attitudes relates to women's 'roles' in society, and in particular women's family responsibilities, and creates an assumption that women are less committed to academia than men. This is, of course, dependent on a particular gendered interpretation of what constitutes commitment.[39] Thus, it has been argued that academic women are disadvantaged because of 'deficit': deficit either by choice, ie lack of commitment, or deficit by not being men.[40] It is perhaps due to the dominance of such assumptions that research suggests that while men find it acceptable to combine family and career in academia, many more women remain childless.[41]

Women on the margins

The combination of some or all of these factors effects the marginalisation of women in the academia. Professor Susan Greenfield has commented that the discrimination which women academics often face is of a most 'sinister' kind.[42] She explains that she does not believe that she has faced direct discrimination, in terms of being denied a job because of her sex, but that she has had to put up with remarks and comments which would not have been made to male colleagues, whilst also being ignored on committees as though invisible, again something which is commonplace for women academics. She concludes that what has to be confronted is how women deal 'minute-by-minute, day-by-day with snide remarks and put-downs'.[43] Evelyn Ellis, professor of law at Birmingham University, similarly comments that she does not believe that she has suffered direct discrimination, but that she has certainly experienced the kind of institutionalised, indirect discrimination encountered by most women.[44]

It is these more subtle forms of discrimination which spring from the pages of research on academic women.[45] The importance of such isolated and insignificant acts is that, individually, they appear minor, and each one may not always be identified as problematic. Accordingly, there often seems to be no safe place from which to question such practices without fear or ridicule.[46] In addition, such isolated minor incidents take much time and energy to ignore or deal with, and cumulatively they may undermine self-confidence and esteem, eventually isolating and marginalising women.[47] Such acts can also be perpetrated with a low probability of detection and recognition. These episodes of discrimination may, therefore, when taken together, have a significant impact on academic women. In other words, acts of discrimination at the margins – the undermining comments, the lack of support, the lack of funds, the additional pastoral roles – may cumulatively undermine the woman, reducing her ability to carry out her work, and so, in the end, 'legitimating' any further decision by her superiors not to advance funds, support or promotion.[48]

This analysis underlines the fact that although the work of an academic is characterised by a considerable degree of autonomy, especially when compared with other forms of employment, it remains a career which is sustained by systems of peer review and professional support from those in senior positions. Thus, advancement can depend on the support of senior members of a department and university who will determine, amongst other things, teaching requirements, nominations for important university committees, research support, sabbatical leave and promotions. Peers in the wider academic community will be responsible for invitations to conferences, to contribute to collections of essays, to collaborate on funding applications and the like. These 'gatekeepers' tend to be men, although slowly there are more women occupying positions of power and influence. And, as in all aspects of public life, the gatekeepers tend to fashion change in their own image.

Feminist legal scholarship

These problems of marginalisation, undervaluation and lack of authority affect all academic women, but further difficulties can arise for the feminist academic woman who draws on her feminism to enrich her practice as a teacher and scholar. Chapter one considered the limited extent to which issues of gender and feminism have infiltrated the law school curriculum, and examined how this may be said to have an adverse effect on the nature of legal education. What might usefully be added to such a discussion is the issue of how the teaching of such courses plays out in terms of prestige within a law school. Are there fewer courses of this nature because the academics who might be interested in teaching them do not wish to be viewed as less authoritative, or mainstream, than their colleagues? Or do academics fear the approbation of their students who may view the teacher and researcher of 'soft' law areas, like gender, as less authoritative or deserving of respect? And, if young, junior, and a woman, does she want to compound her already less authoritative role by taking on this extra burden? The deriding of feminist teaching and scholarship by the student body is well documented and discussed in the North American context, but not in the UK.[49] Is this because there is so little feminist teaching and scholarship by comparison? Or because it is embraced by the student body and academic staff? Or because we fear talking about it or researching it?

In terms of scholarship, it has been argued that the resistance of the legal academy to feminist scholarship is of a different nature to the resistance to many other progressive legal movements, such as critical legal studies.[50] This is because in the academy, feminist scholarship has a 'human face' and because it is 'sustained by a feminist politic that confronts individual men of the academy in light of their practices as colleagues, husbands, lovers and fathers'.[51] The tradition of legal scholarship is 'that which is distant from the scholar may be legitimately studied',[52] with subjectivity and personal experience eschewed. A familiar tactic used to discredit the work of a feminist scholar is, therefore, to label it as personal, political or polemical. Such work is, of course, seen as the 'antithesis of the scholarly ideals of object, neutral and detached, laudations which the man of reason likes to believe characterise his own work and that of his peers'.[53] What is conveniently forgotten is that all scholarship, black-letter or other, is deeply political, with black-letter scholarship being the apotheosis of polemical work in its zeal to maintain and replicate the status quo.[54] Such arguments are echoed by Mary Jane Mossman who notes that feminist legal scholars are expected to think and write using traditional approaches to legal method, and that a feminist scholar who 'chooses instead to ask different questions or to conceptualise the problem in different ways, risks a reputation for incompetence in her legal method as well as lack of recognition for her scholarly (feminist) accomplishment'.[55]

Feminist legal scholarship, broadly conceived, has become more mainstream in recent years; it is published in established journals and by traditional publishers, although it doubtless still carries less weight that non-feminist writing. However, if only certain forms of feminist writing have become accepted, it may be that research on women legal academics remains beyond the pale. It is certainly arguable that some forms of 'feminist' writing have become the 'norm', accepted as 'quality', whereas others, more radical, perhaps more personal, have not.[56] Accordingly, is it that research and writing on women legal academics remains too personal, too political and too risky?[57] There may be a fear that such research would not be thought of as 'proper' research for the purposes of the research assessment exercise (see below) and therefore not useful in terms of peer review. Scholars may be less than keen to put their heads above the parapet for fear of explicit or tacit comments and/or associations, or being 'woman-identified', with a consequently adverse affect on one's career. Accordingly, is it that women legal academics have internalised the 'requirement' that 'quality' research is that which does not declare an interest?[58] Perhaps we should reflect on the reasoning that if feminists are to retain a radical voice, 'we should aim to fit in not *too* well'.[59]

CHALLENGES AND STRATEGIES

Whereas much feminist academic work has sought to document how gender inequality permeates intellectual frameworks and understandings, organisational policy and structures are now ripe for challenge and analysis, in terms of the conditions for the production of knowledge.[60] In particular, the radical changes which have occurred in universities over the last ten to fifteen years have brought new challenges and, perhaps, opportunities for women. Two aspects of these changes will be considered below; the research assessment exercise (RAE) and the equal opportunities agenda.

Assessing research

The accountability agenda of governments over recent years has brought a raft of quality assurance mechanisms to universities, one of the principal initiatives being the RAE. The RAE involves university departments being assessed on the quality of their research, every four to five years, with the next review in 2001. Assessments and gradings are made by a panel of experts appointed by the funding bodies. The rating awarded to each department is a crucial determinant of income and standing within the academic community, with knock-on effects on student and staff recruitment, and the ability of academics to carry out their work. As funding follows quality research, and therefore the prospects of obtaining a job or promotion depend on quality research, the 'publish or perish' or 'write or be written off' clichés have never been more apposite.

The RAE has spawned considerable debate regarding its general effects, although its affect on women academics is rarely mentioned. Reader in Law Joanna Gray moved the debate along when in 1995 she wrote to the *Times Higher Educational Supplement* suggesting certain negative effects of the RAE on women researchers.[61] She referred to the transfer market, the movement between institutions of 'star' researchers for high salaries and other inducements, as a market in 'male academics'.[62] She continued that although there is nothing to stop women playing the transfer market as deftly as men, the problem is that for women, at the stage of their lives when they are most 'marketable', they are often least mobile in terms of domestic responsibilities. It could be added that a job change may result, for many women, in a significant reduction in employment rights at a crucial time, such as maternity leave and pay, which largely remain dependent on two years' service. Having noted that the number of women professors remains (then) at around five per cent, Joanna Gray concluded by arguing that this percentage possibly reflects the number of women who have pursued a traditionally male career path, with limited family responsibilities, and that, unless there is structural change in universities, the representation of women professors is not likely to increase.

A further aspect of the RAE which may adversely affect women, and thereby university departments, is that the census date for the RAE comes every four or five years. Within this time, the researcher has to demonstrate research activity of national or, preferably, international standing. In 1996, there was a requirement to submit four pieces of research, whether they be books, articles, reports or such like, for assessment. The rating of the department, and therefore the university as a whole, depends in law on the performance of individual researchers, and, as noted above, the ratings dictate finances and esteem. For the individual woman researcher, the rating of her work is crucial for professional success and status, including promotion. It can therefore be seen that a woman who has a child during the

RAE period may suffer adversely. Only taking into account the absence from work on maternity leave, this is a considerable inroad into the time available for a woman to undertake her research. Indeed, it is not inconceivable that a woman may have two or three periods of maternity leave within the four-five year census period. Thus, the artificial constriction of the RAE may have an adverse affect on women academics, by measuring their performance regardless of other factors. This means that a department with a predominantly (junior) female staff may be worse off in terms of its rating and therefore in terms of finances, than an all male department.[63] Indirectly, therefore, this may provide an incentive for departments to keep the numbers of their women staff low in order to avoid the potential hazards referred to above.

In order to alleviate the above potential difficulties, it should be possible to note in the relevant RAE documents that a woman has been on maternity leave during the preceding assessment period, and that this should therefore be taken into account. The funding bodies have refused to allow this in previous exercises, but such an change is being sought by the Committee of Heads of Law Schools (CHULS) and the Society of Public Teachers of Law (SPTL), and perhaps by representative bodies of other disciplines. CHULS and the SPTL are seeking an express provision which would enable the impact of disability, long term illness and maternity leave on the work submitted to be considered. If this were done, it would mean that the professional reputation of the individual woman may be less affected by maternity as her RAE 'rating' would take into account the pregnancy and therefore law schools would not be disadvantaged in employing, promoting or rewarding her.

On the more positive side, research into the 1992 RAE found that women had increased their research output to a greater extent than men,[64] which may have a positive effect with regard to the promotion prospects of women academics. Furthermore, an emphasis on research output as the basis for promotion and status, may have the effect of upsetting traditional promotions processes in universities, where preferment came almost as of right to those who had been there long enough. This had the effect of women (and men) having to simply wait their 'turn', women often waiting longer than other colleagues. Thus, the RAE may enable women to 'assert their claims for advancement with greater confidence' and hopefully with greater success, thus disrupting traditional hierarchies.[65] Nonetheless, it must be borne in mind that the assessment of what constitutes 'quality' research is subject to intense negotiation. In the light of the above discussion about feminist legal research, it may be that the positive effects of the RAE are open mainly to those who tread a traditional path in their scholarship.

The equal opportunities agenda

Universities long remained immune from an equal opportunities agenda, basking in the belief that their pursuit of knowledge and truth meant that discrimination within universities was all but impossible. Individual merit and the worth of scholarship were seen (and still are) as neutral concepts on which most right-thinking people would agree. Although universities came in for particular criticism in the government's white paper preceding the Sex Discrimination Act 1975,[66] the situation only began to change when the accountability spotlight was turned on universities.

Part of the government-led requirements for accountability and efficiency, referred to above, have led to demands for equality for underrepresented groups. In response to criticisms, in 1991 the CVCP issued guidance on equal opportunities to all universities.[67] The CVCP also formed the Commission on University Career Opportunity (CUCO) in 1993, the aim of which is to 'encourage and help' universities to 'realise the educational, economic and cultural value of diversity by employing people drawn from all the varied communities which universities serve and influence'. CUCO has produced numerous reports and guidance papers on such things as equal opportunities policies, flexible working, equality targets and childcare.[68] More recently, a guide to *Higher Education and Equality* was published jointly by the CVCP, Commission for Racial Equality (CRE) and the EOC.[69] The national inquiry into higher education, resulting in the Dearing Report in 1998, recommended that institutions should, in the medium term, 'maintain equal opportunities policies' and 'identify and remove barriers which inhibit recruitment and progression for particular groups and monitor and publish their progress towards greater equality of opportunity for all groups'.[70]

At the heart of the above recommendations and initiatives is the equal opportunities policy of the university. Liberal equal opportunity policies, with their emphasis on neutrality and objectivity, merit and excellence, have a particular resonance with the purported values of universities, and indeed the vast majority of universities already have an equal opportunities policy.[71] However, the actual ability of such policies to achieve change is doubted by many. Margaret Thornton has argued that equal opportunities policies only aid those who conform to a stereotypical (masculine) career pattern, and argues that these policies simply perpetuate male hegemony, with the added disadvantage of apparent equality and universalism.[72] A further criticism is that once a policy has been adopted, the institution becomes complacent about equal opportunities, believing that it has done its bit. Allied to this is the fact that often discrimination against women is only seen as direct discrimination, that is, intended, identifiable acts.[73] Thus, little credence or attention is given to institutional and structural forms of discrimination. As a consequence of this, the informal practices and culture of the institution remain the same, with little changing to benefit women.

Although there is some merit in these criticisms, they should be directed at identifying the weaknesses of policies, and aiding their long term effectiveness, rather than being employed as a justification for not introducing a policy. For example, although it may be the case that an institution, having adopted an equal opportunities policy, becomes complacent, this simply points to the fact that, in order to achieve equal opportunities, the adoption of a policy must be seen as only the beginning, with continued pressure and campaigning essential. For example, a 1997 survey by CUCO found that only one third of universities had an action plan to implement their equal opportunities policies.[74] Similarly, as regards the perception of what constitutes discrimination and the continuation of informal practices, the training of individuals and awareness of an appropriately drafted policy may begin to assist in abating such ignorances.

It has, therefore, been argued that equal opportunity policies 'provide a procedural framework and a cultural space within which it becomes possible to take issue with academic convention and practices'.[75] Further, the limited view of equal opportunities policies, described above, represents an 'overdetermined view of women as permanent outsiders'.[76] Thus, charters and committees are only starting points; they do not have to be dead ends. They 'signal a commitment to change, albeit that the direction and force of this has continually to be contested'.[77] Further support for increasing the effectiveness of a policy may come from membership of Opportunity 2000. Campaigning on equality as a business issue, Opportunity 2000 debunks the argument that equality is too expensive.[78] Membership involves setting goals and establishing measurable action programmes to improve women's representation and undertaking a periodic review of progress. In Opportunity 2000 member universities, 15 per cent of vice-chancellors are women, compared with the sector-wide six per cent, and 11 per cent of professors are women, compared with nine per cent in the university sector as a whole.[79] Whether or not this represents the fact that it is the more progressive universities which became members of Opportunity 2000, or whether membership itself has helped initiate change, is not clear.

Thus, the crucial factor is what is done with the policy, rather than the policy itself. The EOC has criticised the lack of any real commitment from universities to equal opportunities, and notes that as well as the simple implementation of equal opportunity policies, the allocation of adequate resources is essential. The EOC also outlines training as an essential element in countervailing the masculine culture of academia, and it recommends that all Senates, Councils and Boards of Governors (the management structures of universities) be required to undertake training on the equality legislation and equality good practice. Furthermore, the EOC recommends that the provision of public funds should be made conditional on such training being undertaken.[80] Thus, in contrast to the woolly words of the Dearing Report,[81] the EOC recommends the establishment of clear targets, incentives and punitive measures in all universities.

CUCO has also recommended that as well as an equal opportunities policy, universities consider 'strengthening maternity provisions, provision of child care facilities, crèches and holiday play schemes and retraining programmes and the provision of flexible working arrangements'.[82] Furthermore, measures are required to change 'attitudes and expectations' in order to 'remove deep seated social and cultural barriers to progression of women, black and minority ethnic groups and people with disabilities'. Although it is not quite clear what some of these recommendations would entail, CUCO recommends 'appropriate support, care regarding promotion, staff development and training, mentoring schemes, networking and the use of role models'.[83]

Thus, the adoption and effective implementation of an equal opportunities policy may begin to exert change, but will only do so when it confronts the culture and attitudes which prevail in an organisation. In order to do so, mere adoption of a policy is insufficient. In addition, any action taken must be directed towards the improvement in the working conditions of all excluded groups.[84]

In considering some of the changes that have taken place in universities, it is important to recognise a paradox. The resistance by many academics to the scrutiny and quality assessments which have been introduced into universities, although having elements of valid criticism, to a certain extent mirror the opposition to exactly the type of scrutiny which feminists and others have sought to establish in response to the monolithic, secretive, and male-dominated inequities of the academy.[85] In other words, it could be argued that the transparency demanded by many feminists is exactly the transparency which may come with greater external assessment and scrutiny. The 'old' regime of higher education was profoundly unwelcoming to women and therefore its passage should not be mourned.[86]

CONCLUSIONS

> 'They will try to persuade you that you are being denied tenure (or promotion or re-appointment) because of your deficiencies. The argument most certain to take you in is the one that speaks to your self-doubt, so they will tell you your publications are mediocre, your teaching weak.
>
> Don't believe it.'
>
> Marcia Lieberman, 1981[87]

This chapter has examined the status of women legal academics, and suggested some of the reasons why the marginalising and undervaluing of academic women in general, and legal academics in particular, may be continuing. These issues are important, not just in terms of the legitimate demand for professional advancement free from discrimination, essential though that is. But they are also significant in terms of the fact that women legal academics play an essential role in determining the culture of the law school and, therefore, the learning environment of law students. An ethos which marginalises women academics and/or feminist scholarship, and in which authority, men and masculinity are inextricably entwined, becomes embedded in students' learning of the law and legal culture. Such an ethos is then perpetuated within legal practice, with an impact on women in the legal profession and on the understandings and interpretations of the law.

The National Committee of Inquiry into Higher Education reported in 1997 that higher education contributes to society by 'adding to the world's store of knowledge and understanding, fostering culture for its own sake, and promoting the values that characterise higher education: respect for evidence; respect for individuals and their views; and the search for truth'.[88] If legal education is to live up to such ideals, steps must be taken to ensure intellectual diversity in law schools, creating a true community of scholars who will add to the store of human knowledge by bringing understandings, perspectives and experiences from a wide range of interests and backgrounds.

PERSONAL TESTIMONIES OF WOMEN LEGAL ACADEMICS

Professor Jo Shaw, Faculty of Law, University of Leeds

Jo Shaw is Director of the Centre for the Study of Law in Europe at Leeds University. She describes below the development of her career, her views on the changing nature of higher education and on being a woman legal academic ...

'I come from a rather solid northern middle class background, attending a Girls' Grammar School. My father was a solicitor and my mother taught modern languages. Hence, I must subconsciously have been trying to please both of them when I was interviewed at Cambridge in Modern Languages, with a view to changing after one year to do law. I would not call my school desperately 'pushy' in the ambitions for its girls, but those who expressed a desire to try Oxbridge entrance were well supported. My brother was just finishing at St Catharine's and he advised me to apply to

Trinity on the grounds that it was 'big and rich'. These indeed proved to be its defining characteristics, as well as being full of mathematicians and scientists. I duly changed from modern languages, and graduated with a solid run of Upper Seconds in both parts of tripos. Some time during my third year, I asked one of the fellows what he thought about an academic career for me. He strongly advised against. I have ignored his advice.

In the meantime, I was selected by some mysterious process (old boy's network?) for a scholarship to take a Master's degree in European Law at the Institut des Etudes Européennes in Brussels. I consequently deferred the statutory place at the College of Law and articles in the City, which I had duly obtained like virtually all my peers – although it has to be said that I was not overburdened with offers, for whatever reason. The experience of living abroad was instructive, and although the course was not desperately academic, continuing to study reinforced my feeling of dissatisfaction with the idea of working as a solicitor. I applied for the one academic job I saw advertised in *The Times*, namely a research assistantship in EC law at University College London and was successful. Again, I deferred my articles and place at the College of Law – this time to the increasing impatience of my parents. Finally, after about six months I withdrew completely – acknowledging that a career in the City was not for me.

I could have stayed longer than one year at UCL, but I started applying for lecturing jobs and was successful reasonably quickly in securing a post at Exeter University, initially focusing on German Law. Since then, I suppose one would describe my career development as having been surprisingly smooth. Six years followed at Exeter, moving slowly up what became the Lecturer A scale, until I was plucked from obscurity and caught on the waves of market forces crashing against the bulwarks of tradition and convention in the universities as the impact of the RAE began to gather strength. I was invited to apply for a much better job at Keele University. I stayed there for five years, leaving there as a Reader to take up a Chair in European Law at the University of Leeds in 1995. I am Director of the Centre for the Study of Law in Europe and have been responsible for developing the research profile of the Department in the general 'European' domain. I am now lucky enough to work with many women staff and postgraduate students; I have to say, it makes a difference.

It is quite apparent to me that over the years much of legal academia has become increasingly feminised. When I first joined the SPTL, I was invited to bring my 'wife' to the Annual Conference. While it was an interesting and thought-provoking proposition, I found it shocking that a professional association could make such assumptions in the mid-1980s and did not hesitate to write and tell them so. Certainly legal academia has become – at least outwardly – a place more tolerant of different lifestyles, personal attitudes and even sexual orientations. Since I began work, quite a number of women have moved into senior positions in the universities, and a whole host have joined the junior ranks. The upper ranks remain male-dominated to large extent, nonetheless. I think the acid test is that I have never been in a position to use a woman as a referee (although, I ask myself on re-reading this, would I have done so at that time, if it had been possible?)

In addition, legal academia has offered an environment in which many (although undoubtedly not all) gay men and lesbians have felt able to come out and the extent to which such value changes have become properly embedded rather than being merely a veneer of liberal attitudes has become increasingly important to me. Sadly, there remain very few black, Asian or other ethnic minority staff in law schools, although in some ways law schools are quite multicultural because most have a number of staff from overseas. While I have personally encountered little direct hostility to my career progress, I have nonetheless occasionally detected undercurrents of uneasiness, mainly, but not solely, on the part of male colleagues. However, it is possible that these could be associated as much with the dramatic changes in universities in the last ten years as they are with me as a senior academic woman. For me, not all of those changes have been a bad thing, and it strikes me that universities are now in general terms (although not always and not for everyone) better places for women.

In many respects, working in a university has offered me the ideal career. I had one son, at a relatively early age, whom I now bring up as a single parent. Universities had maternity leave schemes in place when they were still a twinkling of the eye in most areas of legal practice. Over the years, I have been able to take advantage of the flexibility of working as an academic which legal practice would not have permitted me. That said, I would say that success as an academic requires you to work just as hard as you would do in practice, but in rather different ways. I have also been lucky enough to be doing things which, on the whole, I feel confident about doing, namely teaching and writing; thinking up and taking forward new ideas for academic development; and, from time to time, money making. I regret to say that like many of my colleagues I spend far too much time these days pushing papers around and filling in pointless forms. I have been supported in my work by a whole host of friends and family. I have also benefited from working in a number of genuine academic 'communities', where the boundaries between colleague and friend, and talk between friends and fruitful academic discussion have rapidly melted away.

The career path of a legal academic is now much clearer than it was more than fifteen years ago when I first contemplated it. Although there is pitifully little financial support available, it is nonetheless generally expected that recruits to academic jobs will now have research degrees and even publications before they start. Often that means going abroad for students from the UK, since there appear to be more scholarships to leave the country than to stay. Since the end of the law school boom in the late 1980s, the number of vacancies has become fewer, although the inevitable mobility between academia and practice (and other careers such as work in the EU institutions) means that there will always be at least a trickle of jobs (and a trickle which looks like a flood when it is compared to other social sciences or the humanities). Organisations such as the Socio-Legal Studies Association (SLSA) run postgraduate conferences which include valuable advice sessions on 'getting an academic job'. More short term or rolling contracts are used, and this can create great insecurity and uncertainty. Once in, there are greater pressures to publish from the very beginning and probation has become a meaningful hurdle, but conversely new academic staff enjoy support systems through courses on teaching and learning, and mentoring and review systems which were largely unknown until the concept of 'quality management' hit universities. I feel strongly about fostering the career development of new academic staff, and see it as very much part of my job description.

Universities are less isolated places than they were – especially, I think, for women. Communication systems as varied as e-mail and an active conference and seminar circuit keep people across the different universities and different countries in touch with each other. The networking fantasy of the 1970s feminist is now a (cyber) reality for us all. Conferences provide opportunities to develop ideas before an audience which were lacking when I began. They are fora where women are ever more visible, although I have to say that I find the SPTL to be lagging far behind the SLSA in this respect. And all the time, the ever more pervasive effects of market forces in the universities provide both opportunities for preferred advancement for those who are lucky or whose faces fit, as well as pitfalls for those whose faces do not, or who fall foul of unexpected efficiency drives.

Readers will see that I came into legal academia from a very conventional academic background, and that I have probably benefited from my independent schooling and Oxbridge education in ways which have smoothed my path, even as a woman and a woman who has always had quite decided views about many things and has never hesitated to describe herself as a feminist. Formative periods of my career have coincided happily with the opening up of opportunities in law schools, and I have had the independence and means to move around the country even with a small child to take up those opportunities. To that extent, I am a bad (untypical?) person to give advice about coming into, or staying out of, legal academia. However, my view remains that it is a rewarding, if surprisingly tough, world to be involved in. The financial rewards are notoriously small in comparison to legal practice (although typically greater than those achieved by one's colleagues in equivalent positions in other departments), but there remain other compensations. There is little to match the satisfaction of first seeing one's name in print.'

Professor Katherine O'Donovan, Faculty of Laws, Queen Mary & Westfield College, University of London

Katherine O'Donovan has written many pioneering works in the field of feminist legal studies and describes below her journey to become one of the few women law professors in the UK. She gives her views on how things have improved for women, and offers some advice on becoming a legal academic ...

'My overriding need, as a child and student, was to become financially independent. My mother's dependence on her 'housekeeping money' and my father's sense of obligation both to maintain a middle class household of six children, and to spend, made for an uncomfortable home life. School and university fees, dentists' and doctors' bills, mounted, as the Ireland of the fifties did not have a welfare state. But something else was happening. In retrospect, it is evident that a movement of young women into professional life took place globally in the sixties.

Patriarchy ruled at home, and as I had studied modern languages at school, my father decreed that I should do so at University College Dublin with a view to becoming a secretary. To me, my studies were a continuation of the boredom of school. Some of this must have dawned at home, for after Christmas I was permitted to change to law, of which I had talked for years, despite my father's forebodings that it would lead to the poorhouse. His views were based on the career of one of his brothers of whom it was claimed that he had not earned a penny at the Bar until he was thirty-five. As a result, the Bar or being a solicitor's apprentice, both of which cost fairly large sums of money, were beyond my scope.

During my five years of legal studies, in Dublin, and on scholarships in Strasbourg and Cambridge, Mass., no woman taught me, nor did I ever meet one who could have. But then, even in the enlightened USA, only five per cent of the law students were women. On my arrival in 1965 at the Queen's University of Belfast as an assistant lecturer, I felt grateful for my membership of the academic and male club. The ethnic and gender stereotyping, which was endemic in the daily coffee time conversation, seemed a small price to pay.

As it turned out, things were on the move, in Northern Ireland and in the women's movement. A language in which to describe and to object to gender and racial insults very gradually developed, and is now an unnoticed part of everyday usage. In the meantime, for the sake of adventure, I went off to Africa: the Haile Sellassie I University in Addis Ababa.

The above account of my early years as an academic is intended to make the point that changes in attitudes and expectations are part of both general and individual histories. Pinning down all the factors is very difficult. The women entering the legal profession today may also find that their training period is unpaid, or that they cannot survive at the Bar. It would not surprise me to hear that the custom of requiring a fee for a training place is revived.

From Addis Ababa, I moved to Sussex on a temporary post and then on to Kent. A great breakthrough came when a group of women managed to establish an interdisciplinary course on 'women in society' at Kent in 1975. We were open to ridicule, and accused of teaching knitting. I learned an enormous amount from the other members of the team in sociology, anthropology, social policy and economics. I read beyond the feminist bestsellers. In the writings of the women's movement, I found a self who had been long suppressed. It is difficult now to convey the sense that many of us had in reading that, 'this is about me'. We had not found our perspectives in print previously. Much legal writing has now been published on women and gender and this silence has been filled with voices. Then, the realisation of a silence was itself a shock. At that time only writings of a 'black-letter' nature, or in the positivistic empirical tradition were regarded as respectable.

The first women law professor had been appointed in the UK in 1971, overcoming many obstacles. My recollection is that 'academic respectability' was important to the early women law professors and, and having faced the difficulties of assimilating themselves to the male pattern of success, they were not 'women-identified'. Nevertheless, I received a good deal of encouragement and support from Clare Palley and Rosalyn Higgins, both of whom were professors at Kent in the seventies. A group of Women Law Teachers, a women only group, was formed in the early eighties which was highly supportive. Our purpose was to hold workshops and encourage the establishment of courses which permitted a feminist perspective. But we also enjoyed comparing notes on the homocentricism of legal materials, judges and our colleagues.

As is clear from the above, the sixties and seventies were a time of reassessment politically. The left initially rejected a gender perspective, insisting that stratification should only be seen in class terms. This meant that those male colleagues from whom one might have expected sympathy were fighting their own battles. This can be confirmed by turning to the British Journal of Law and Society (now JLS) which first appeared in 1974 as a radical journal. The first six volumes dealt with theorists of the sociology of law, with criminal justice and with industry. There is a silence about feminism. Gender appears only briefly as a category of criminal defendants. Empirical legal work was radical and it is noticeable that when early British legal work on women appeared it was in the empirical tradition. In a sense feminists were caught between a need to establish their respectability as academics and their anger at the rules which precluded writing about women's silence and the injustice of laws and legal processes. Because the personal is political such writings could not be 'objective'. This paradox has been resolved in the nineties with much interesting work emanating from writers on gender and queer theory, who are often writing about themselves.

My own progress in career terms was helped by my finding a voice in which to express the anger of women about the law. Gender neutrality, later to be criticised as a goal, did not exist in much of legislation and case law. The strategy was to confront liberals with liberal arguments about equality. Despite all the criticisms of this as narrow and gradualist, the idea of other perspectives emerged. As a result the conceptions of self, and possibilities of becoming and being, open to women today, are much greater than when I was young. The apogee of my ambition on arriving in Britain was to reach the level of Reader in a University. It never occurred to me that I would be a professor in my forties. Admittedly, admission to that august title is inevitably diminished by the arrival of women, or so it seems.

In summing up my experiences of a career as a legal academic, I return again to ways of knowing and being. The feminist challenge is to traditional analysis and perspectives. It has enabled new ideas to be taken up and to enter official discourse. For those who saw feminism as political beyond academia this is a source of satisfaction. But above all what strikes me is the ambition of my women students whose sense of self is not 'cribbed, cabined, and confined'. Diverse voices have broken the idea that women can be discussed as category. However, despite the fragmentation that has resulted from a multiplicity of voices, the breaking of silences and the resulting pluralism have created richer and more diverse worlds in which we can all share.

My advice to women thinking about an academic career is to write. My opinion, which is probably a minority view, is that one should write about issues that one finds interesting and not for the sake of academic respectability. This is a job in which you can travel. I spent a year in Florence, two years in Malaysia and three in Hong Kong. Visiting and going to conferences in all corners of the globe is a common experience. Universities increasingly accept working from home, although it is advisable to inquire about this at interview. Academics are individualists in a similar position to the self-employed. As a result one's colleagues are a diverse bunch amongst whom one can find kindred spirits. In addition the contact with students is refreshing and, in my case, is a welcome antidote to intellectual demise.'

Professor Evelyn Ellis, Faculty of Law, University of Birmingham

Evelyn Ellis has worked for most of her academic life at Birmingham University, teaching and researching in the fields of discrimination law and European law. She describes below her experiences, and offers some advice for would-be academics ...

'I was born and brought up in London and attended St Paul's Girls' School. Pupils were encouraged to aim high and it was not countenanced that we would face discrimination or other obstacles in later life on account of our sex. When it came to choosing A-level subjects, it was not really clear whether I was a scientist or an artist. I chose science subjects because I believed that otherwise I would never have a proper understanding of things scientific. As it turned out, I still do not understand things scientific and my A-level grades are best forgotten! However, I did come to realise that I wanted to study a subject which combined logic with a love of words and politics. Law seemed the obvious answer.

I went on to study law at Girton College, Cambridge, at that time still an all-female establishment. Cambridge was my first exposure to a male-dominated world; our class consisted of something in the region of eight women students and 200 men. I found that imbalance daunting and was conscious of being branded a blue-stocking. Although the legal education I received at Cambridge was superb, I did not find the overall environment particularly comfortable or welcoming.

After graduating with a first, I toyed with the idea of a career at the Bar, but ended up choosing academic life, largely because it seemed to offer more flexibility. I then went on to study for an LLM at Birmingham University and thereafter took the Bar Exams. I was awarded a PhD by Birmingham in 1989.

I have worked almost all my life at Birmingham University, progressing from lecturer in 1972 to Professor of Public Law in 1996. I spent two years teaching at Boston College, USA, in the mid-1970s, when my husband was studying for the MBA at Harvard. My research has centred on EC Law and discrimination law, two areas which have fortuitously come together and achieved a high profile in recent times.

I do not believe that I have suffered direct discrimination in my working career, but I have certainly experienced the kind of institutionalised, indirect discrimination encountered by most women. In particular, I have had to work extremely hard to bring up my family, at the same time as keeping up my career because, despite the support of a generally sympathetic husband, I have found myself bearing the bulk of domestic responsibilities. In addition, because there have been two careers in our family, I have not been able to move around from job to job and thus have taken longer than I otherwise might have to achieve promotion. On the plus side, I have gained from this a solid base of local friends and a feeling of rootedness to my community.

One of the most irritating experiences is constantly hearing the comment that you have got a perfect job because you do not have to work during university vacations. In fact of course research output is the critical factor influencing an academic career and one of the most difficult aspects of the job is managing to engage in concentrated research when the family is at home on holiday; not only is it physically difficult to find time and energy for research but I have found that this is when guilt about not being a 'full-time mother' weighs most heavily upon me. Women in general do not progress well within universities; they remain concentrated at the lower levels of the academic hierarchy and this is largely, I believe, because of their relatively poor research output.

My advice is therefore this. First, organise your time with military precision. Remember that your male colleagues waste plenty of time boozing and gossiping and remind yourself that time with your children, if you have any, is both better spent and more enjoyable. Secondly, make sure that you carve out a research niche for yourself and get published regularly and in the right places. Thirdly, never be tempted to become a part-timer. If you do, you will work just as hard as your male colleagues but for less money and you will never again get properly back onto the career ladder.

Universities today are acutely aware of the underrepresentation of women staff and many are taking active steps to address the problem. If you are genuinely interested in the intellectual discipline of law, want a career which will allow you the luxury of being able to worry away at a problem until you have solved it, and possess the self-discipline necessary to force yourself to sit down to write an article in midsummer, then an academic career is certainly open to you.'

Ruth Deech, Principal, St Anne's College, Oxford

Ruth Deech began her career as a legal academic and is now Principal of St Anne's College, Oxford and chair of the Human Embryology and Fertilisation Authority. She describes below her experiences of the law and academia and considers the future prospects for women in the law ...

'My father was a refugee from Vienna, where he had studied law but had not been allowed to practice by the Nazis. He told me that he had enjoyed his law studies but had eventually been drawn into journalism as a means of self-defence for him and other persecuted Jews of the period. He arrived in this country on 3 September 1939 with no belongings at all, no English and few prospects. He met and married my mother here and they settled in South London. I attended a local primary school, and at the age of ten I won a scholarship to Christ's Hospital, Hertford. I went to study law at St Anne's College, Oxford, leading an average undergraduate career, with a great deal of social life and travel, but I enjoyed law and ended up with a first class degree. I subsequently went to Brandeis University in Boston and took an MA in Contemporary Jewish Studies. I did Bar finals, but I believe that my desire to practice law was already fading by then.

I took up teaching law when I went with my husband, who was then a physicist, to the University of Windsor, Canada. I loved it from the first moment, both the theatrical side of appearing as a young lecturer before a large class of confident and articulate north Americans, and also the personal involvement that grows from Oxford tutorials. I was well taught at Oxford and started tutorial teaching there on a locum basis almost as soon as I had graduated. After twenty-one years as a tutor at St Anne's College, and many years holding senior administrative positions within the College and in the University, I was elected Principal in 1991. I have now stopped teaching law, which I miss, although I give the occasional lecture. Most of all, I miss the excitement of involvement with a student's choice of career and their progress and it gives me special pleasure to keep in touch with the graduates and follow their careers.

Family law was my speciality and formed a valuable background when I was offered the chairmanship of the Human Fertilisation and Embryology Authority (HFEA) in 1994. Chairmanship of the HFEA, although stressful because of the nature of the issues under scrutiny by the media, combines two facets of law that I find particularly fulfilling. One is the human element: the pleasure of knowing that some of the desperately unhappy, infertile couples (it is said that one in six of British couples seek help with infertility problems) will have healthy babies. All our work is guided by certain ethical principles, namely, safety first, and the avoidance of risky, untried treatments, the welfare of the potential child, and human dignity and autonomy. At the same time I enjoy chairing a committee of twenty-one able and committed members, trying to resolve difficult legal and social problems. Participation in the Diane Blood case was especially instructive and at the end I rather regretted my decision not to practise at the Bar.

At the Inner Temple in the early sixties, I was too inhibited about going forward. I believed that I had no contacts and did not have the money to enable me to take up a career at the Bar. Looking back, I can see that I lacked the enthusiasm to overcome those difficulties and that they were more imaginary than real. I had been too shy to make the effort necessary to come to know barristers at the Inn and I had not inquired thoroughly into the financial possibilities of life at the Bar. Nevertheless, there were real difficulties. I recall that in the early 1970s many of my women students of high quality would seek articles with well-known London solicitors' firms and would be rejected on the sole ground that they were women. This was before the Sex Discrimination Act of 1975, and the rejecting firms did not try to disguise the reasons for rejection. I saw, and have kept, letters from firms saying to students: 'We have a quota on female students'; or 'We do not take women'. I took up the cudgels on their behalf but it was only when the law changed that they were able to find articles.

The other problem was money. Many a bright student told me that he or she could not afford to go to the Bar because they could not keep themselves during pupillage and faced the prospect of uncertain tenancy. This was compounded when a student came from a modest or broken home.

Once Chambers started to pay pupils, the situation was transformed and the most talented and suitable pupils went to the Bar, rather than to solicitors' firms because of the pay. I cannot emphasise strongly enough the changes wrought in recruitment by the ending of discrimination and the payment of a living wage to pupils, and the sense of satisfaction that this brought me.'

Dr Hilary Sommerlad, Senior Lecturer in Law, Leeds Metropolitan University

Hilary Sommerlad gained her PhD in politics, qualified as a solicitor and practised in a small firm dealing with many race and civil liberties issues. She has had a varied career in legal practice and is now an academic. She describes her experiences below...

'I was born in Oxford and brought up on a council estate on the edge of the city. I was the youngest of three children, and my brother and I were the first members of the family to get to university. I had been told by my head teacher that I wasn't University material. I was determined to show her she was wrong and went in for Cambridge, where I studied History. I hadn't considered law.

After Cambridge, I would very much have liked to go to the Bar, and I wanted to work in an area like immigration law. However, I'd already found Cambridge difficult in the sense that the culture was very powerfully middle class, and I felt an outsider. There were also very strong messages around then that law was not a career for women. I remember a male friend commenting that it would 'make me hard'. But finally, my family could not possibly have supported me through the expense of the College of Law, Bar exams and pupillage.

On the off chance, I replied to an advertisement seeking PhD candidates in Politics at York University. I had wanted to do research into the Spanish Civil War, but subsequently decided to look at theories of underdevelopment in the context of a nineteenth century war between Peru, Bolivia and Chile. Even though the work was so distant, geographically and in terms of time, it involved me in looking at theories of the State, and experiencing life in a country where the idea of the rule of law was more tenuous, and more explicitly a weapon in the hands of the powerful. As a result, and after a particularly interesting research seminar in York on international law, the idea of retraining in the law returned.

I felt that I wanted a career which would be both secure and of real use to someone. I had to teach part-time whilst taking the (full-time) CPE, and then take a year out teaching and doing contract research. I was then articled to Howard Cohen's in Leeds. Although it was a general practice, it attracted some very interesting work. As a result, I worked on some extremely interesting trials which involved important race and civil liberties issues. I was also able to see the law used in a very creative and innovative way to address significant issues, and watch some excellent barristers from radical London chambers in action. I was also very lucky in that Howard Cohen was an excellent lawyer who treated me with complete respect and believed in my abilities. He had a real insight into the experiences of marginal groups, and in common with many other Jewish firms in Leeds, believed that a core aspect of being a lawyer was to use the law to help these groups.

After finals, I went into general practice, but I was beginning to feel that it was difficult to find an environment where you could practice without the work consuming all aspects of your life. In particular, the areas of law that I found most intellectually exciting, like international law, would have involved a very disrupted private life, particularly since I felt that a woman of my age would have had to work that much harder in order to prove herself. I had been teaching part-time at Leeds Metropolitan University for a while, and when a full-time job came up, I took it. Although I missed practice, I welcomed the opportunity to use my teaching skills and to develop my intellectual understanding of law.

When I took up the post, I was asked to take over responsibility for a project on women returners to the solicitors' profession. The project was commissioned by the Law Society, and involved surveying and interviewing local women solicitors. I'd taught feminist theory, and so I wasn't surprised by the way in which the research underlined the institutional and cultural barriers which women confront in attempting a career in the law. I'd consistently met them myself, even in the academia, which one might expect to be a more 'friendly' environment. I felt, however, that it was an area which really did merit far more attention and systematic explanation. This feeling was reinforced by my own experiences of trying to maintain and develop a career after the birth of my child. As a result I've carried out more research on this theme sponsored by the Law Society, which has been the substance of several publications including a book entitled 'Gender, Choice and Commitment'. My work in this area points to the importance of general cultural change which will allow both women and men to balance their private and public worlds: until this occurs, women's participation is always likely to be at considerable personal cost.'

NOTES

1   Hansard Society, *Women at the Top*, Hansard Society: London, 1990, p 68.
2   *The Subjection of Women*, 1869, discussed in Martha Nussbaum, *Cultivating Humanity – A Classical Defense of Reform in Liberal Education*, Harvard University Press: Harvard, 1997, pp 186–189.
3   Although it is not suggested that only women can write and research about women.

4   Karen Gold, '8.1 per cent of professors are women', *Times Higher Educational Supplement*, 3 July 1998. A slight increase from the 7.3 per cent in 1994–95 (Sian Griffiths, 'Chipping away at the glass ceiling', *Times Higher Educational Supplement*, 26 July 1996), and the three per cent in 1989–90 (AUT, *Sex Discrimination in Universities*, 1992, p 2).

5   *Times Higher Educational Supplement*, 5 June 1998.

6   AUT Press Release, 'Unfairness at work in academe: women coming off worse', 15 May 1998.

7   Quoted in Lola Young, 'The colour of ivory towers', *Times Higher Educational Supplement*, 5 June 1998.

8   Vice-chancellors (Principals in Scotland) are equivalent to the chief executive or managing director of a company: Maureen O'Connor, 'Self help for women on the way to the top', *Times Higher Educational Supplement*, 25 June 1998.

9   AUT, supra note 4, p 2.

10   Robert McNabb and Victoria Wass, 'Male-female Salary Differentials in British Universities', 49 (1997) *Oxford Economic Papers* 328, p 328.

11   John Wilson, 'A Survey of Legal Education in the UK', 9 (1996–97) *Journal of the Society of Public Teachers of Law* 1, p 5.

12   See John Wilson, 'A third survey of university legal education in the UK', 13 (1993) *Legal Studies* 143–182; Phil Harris and Steve Bellerby, *A Survey of Law Teaching 1993*, Sweet & Maxwell and the ALT, 1993; and Phil Harris and Martin Jones, 'A Survey of Law Schools in the UK 1996', 31 (1997) *Law Teacher* 38–126.

13   The up-dating surveys only disaggregate student statistics according to sex.

14   HESA keeps records relating to 'cost centres' which are not equivalent to law schools.

15   On suggestions for a research agenda on women legal academics, see Fiona Cownie, 'Women Legal Academics: A New Research Agenda?', 25 (1998) *Journal of Law and Society* 102–115.

16   Full details of the methodology and returns for each law school can be found in appendix three.

17   See appendix three for details of the methodology employed in the survey and for details of those universities which responded.

18   The universities of Liverpool and Glasgow.

19   Symposium, 'Black Women Law Professors: Building a Community at the Intersection of Race and Gender', 6 (1990–91) *Berkeley Women's Law Journal* 1–201, p 1.

20   Margaret Thornton, 'Discord in the Legal Academy: the case of the feminist scholar', 3 (1994) *Australian Feminist Law Journal* 53 and *Dissonance and Distrust: Women in the Legal Profession*, Oxford University Press: Oxford, 1996, pp 106–129.

21   This is not to deny that some men are outsiders in the academy. However, as has been argued in articulating a male typology, the subject positions open to men are generally empowering. See Richard Collier, ' "Nutty Professors", "Men in Suits" and "New Entrepreneurs": Corporeality, Subjectivity and Change in the Law School and Legal Practice', 7 (1998) *Social and Legal Studies* 27–53.

22   A point made by Elizabeth Schneider, 'Task Force Reports on Women in the Courts: The Challenge for Legal Education', 38 (1988) *Journal of Legal Education* 87–99.

23   'Discord in the Legal Academy', supra note 20, p 57.
24   Ibid p 53.
25   Supra note 4, p 8.
26   Supra note 10, p 340. This was confirmed by an analysis of starting salaries, with women being paid on average 6 per cent less than men.
27   Ibid p 338.
28   Patricia Leigh2ton, Tom Mortimer and Nicola Whatley, *Today's Law Teachers: Lawyers or Academics?*, Cavendish: London, 1995. This research was based on just over 1,000 completed questionnaires from legal academics in all sectors of higher education, 42 per cent of whom were women.
29   Ibid p x.
30   Ibid p 71.
31   CVCP, *Promoting People: a strategic framework for the management and development of staff in UK universities*, London, 1993.
32   *Dissonance and Distrust*, supra note 20, p 116.
33   EOC, *Response of the EOC to the National Committee on Inquiry into Higher Education*, 1996, para. 26.1.
34   Supra note 4, p 13.
35   Supra note 1, p 66.
36   Ibid p 68.
37   Helena Kennedy, 'Introduction', in Sian Griffiths (ed), *Beyond the Glass Ceiling – Forty women whose ideas shape the modern world*, Manchester University Press: Manchester, 1996, pp 1–9, p 4; and Helena Kennedy, 'Who's been sitting in my chair?', *The Guardian*, 8 July 1996.
38   Jackie West and Kate Lyon, 'The Trouble with Equal Opportunities: the case of women academics', 7 (1995) *Gender and Education* 51–68, p 62.
39   See further chapter four under the heading 'A baby bar?'.
40   Sandra Acker, 'Women: the other academics', 1 (1980) *British Journal of the Sociology of Education* 81–91.
41   Pauline Leonard and Danusia Malina, 'Caught Between Two Worlds: Mothers as Academics', in Sue Davues et al (eds), *Changing the Subject – Women in Higher Education*, Taylor & Francis: London, 1994, pp 29–41.
42   Quoted in Judith Judd, 'Bias that stops women academics reaching the top', *The Independent*, 7 June 1997.
43   Ibid.
44   See her personal testimony at the end of this chapter.
45   See, for example, Louise Morley, 'Glass Ceiling or Iron Cage: Women in UK Academia', 1 (1994) *Gender, Work and Organization* 194; Louise Morley and Val Walsh (eds), *Feminist Academics: Creative Agents for Change*, Taylor & Francis: London, 1995; Ann Brooks, *Academic Women*, Open University Press: Buckingham, 1997; Sue Davues et al (eds), supra note 41; Sian Griffiths (ed), supra note 37; Miriam David and Diana Woodward (eds), *Negotiating the Glass Ceiling – careers of senior women in the academic world*, Falmer Press: London, 1998.
46   Louise Morley, ibid p 197.
47   Barbara Bagilhole, 'How to keep a good woman down: an investigation of the role of institutional factors in the process of discrimination against women academics', 14 (1993) *British Journal of the Sociology of Education* 261–274, p 270.

48   See also chapter four under the heading 'Discrimination at the margins'.

49   See Mona Harrington, *Women Lawyers – Rewriting the Rules*, Plume-Penguin: USA, 1995, pp 41–68; Ruth Colker, *Pregnant Men – Practice, theory and the Law*, Indiana University Press: USA, 1994, pp 41–70; and the discussion of some Canadian experiences in Fiona Cownie, supra note 15.

50   For an entry into the considerable US research on attitudes to feminism in the academy, see Veve Clark et al (eds), *Anti-Feminism in the Academy*, Routledge: London, 1996.

51   'Discord in the Legal Academy', supra note 20, p 57.

52   Ibid.

53   Margaret Thornton, 'Hegemonic Masculinity and the Academy', 17 (1989) *International Journal of the Sociology of Law* 115–130, p 125.

54   Ibid pp 124–125.

55   Mary Jane Mossman, 'Feminism and Legal Method: The Difference it Makes', in Martha Albertson Fineman and Nancy Sweet Thomadsen (eds), *At the Boundaries of Law – Feminism and Legal Theory*, Routledge: London, 1991, p 297.

56   Such as writings on lesbianism or cultural feminism: see Andrea Loux, 'Idols and Icons: Catherine Mackinnon and Freedom of Expression in North America' and Rosemary Auchmuty's review of *Sourcebook on Feminist Jurisprudence*, in 6 (1998) *Feminist Legal Studies* 85–104, 135–137, respectively, who both consider the dominance of certain forms of feminist legal thinking.

57   This is not to say that there is no such writing, just that it is scarce. See for example the personal testimonies of women legal academics at the end of this chapter and Anne Bottomley, 'Feminism, the Desire for Theory and the Use of Law', extracted in Hilaire Barnett (ed), *Sourcebook on Feminist Jurisprudence*, Cavendish: London, 1997, pp 84–91. There is also a 'lack of self-reflection' in writing about masculinity in law and the sexual politics of academic practices: see Richard Collier, 'Masculinism, Law and Law Teaching', 19 (1989) *International Journal of the Sociology of Law* 427–451, p 437.

58   It is not suggested that this describes all writers. For example, Sandra Fredman comments in the preface to her book *Women and the Law* that: 'I come to the topic of Women and the Law without attempting to conceal my passionate concern for the subject.' She goes on to state that in writing the book she will 'draw explicitly on the personal as well as the academic': Sandra Fredman, *Women and the Law*, Oxford University Press: Oxford, 1997, p ix.

59   'Discord in the Legal Academy', supra note 20, p 71, emphasis in original.

60   See further Louise Morley and Val Walsh, 'Feminist Academics: Creative Agents for Change', in Louise Morley and Val Walsh (eds), supra, note 45, p 2.

61   *Times Higher Educational Supplement*, 17 November 1995.

62   Although research on the 1992 RAE suggests that the existence of a transfer market is a 'myth' (Ian McNay, 'The Impact of the 1992 RAE on Individual and Institutional Behaviour in English Higher Education', Centre for Higher Education Management, Anglia Polytechnic University, 1997), there may have been more movement in 1996, and indeed it is a self-fulfilling prophesy, and certainly played out in the actions and attitudes of departments.

63   Although clearly not all women wish to or can have children and not all women are of childbearing age.

64   Supra note 62, p 17.

65   Joanna de Groot, 'After the Ivory Tower: Gender, commodification and the "academic"', 55 (1997) *Feminist Review* 130–142, p 138.

66   *Equality for Women*, Cmnd 5724, 1974, paras 11, 13–14.

67   This document replaced a 'code of practice' which was viewed as too threatening to university autonomy: Christine Heward, 'Academic Snakes and Ladders: reconceptualising the "glass ceiling"', 6 (1994) *Gender and Education* 249–262, p 259.

68   The following are available from CUCO (contact details in appendix five) for £5.00 each: *Childcare in universities and colleges*, 1996; *Equality targets – action planning and monitoring in universities and colleges*, 1996; *Flexible working in universities and higher education colleges*, 1997; *A report on policies and practices on equal opportunities in employment in universities and colleges of higher education*, 1997.

69   This comprehensive guide covers all aspects of higher education, including recruitment (students and staff), teaching, research, quality assurance and university management. It sets out the clear case for equal opportunity measures and gives examples of good practice and is available from the EOC for £5.00.

70   The National Committee of Inquiry into Higher Education, *Higher Education in a Learning Society*, 1997, recommendation 49. This recommendation does not, however, appear in the summary report, suggesting that it was not considered to be a high priority.

71   A 1997 survey by CUCO, *A report on policies and practices on equal opportunities in universities and colleges in higher education*, revealed that 96 per cent of universities had an equal opportunities policy.

72   Supra note 53.

73   Supra note 38, p 60.

74   Supra note 71.

75   Supra note 38, p 66.

76   Ibid.

77   Ibid.

78   Although reliance on the economic case may be problematic: see further Clare McGlynn, 'The Business of Equality' in Clare McGlynn (ed), *Legal Feminisms: theory and practice*, Dartmouth-Ashgate: Aldershot, 1998.

79   Opportunity 2000, *Fifth Year Review*, Opportunity 2000: London, 1996.

80   EOC, *Response of the EOC to the National Committee of Inquiry into Higher Education*, EOC: Manchester, 1996, para 22.1.

81   Supra note 709.

82   CUCO, *Response from CUCO to National Committee of Inquiry into Higher Education*, 1996, para 11. See also the AUT's, *Equal Opportunities, Employment and domestic responsibilities*, 1997.

83   CUCO, ibid para 12.

84   The AUT announced in 1998 an initiative to include sexual orientation discrimination within equal opportunities policies: see AUT, *Pride not prejudice – equal opportunities in higher education: a lesbian, gay and bisexual perspective*, 1998. In addition, the CVCP, CRE, CUCO, NATFE and the AUT are sponsoring research into the under-representation of ethnic minorities in academia.

85   Supra note 65, p 139.

86   Celia Davies and Penny Holloway, 'Troubling Transformations: Gender Regimes and Organizational Culture in the Academy', in Louise Morley and Val Walsh (eds), supra note 45, p 8.

87   Marcia Lieberman, 'The most important thing for you to know', in G Desole and L Hoffman (eds), *Rocking the Boat: Academic Women and Academic Processes*, Modern Languages Association of America: New York, 1981, p 3.

88   Supra note 70.

Chapter 3
In training: the Legal Practice Course and training contracts

'There are many employments in life which are unfit for the female character. The profession of the law must surely be one of them. The peculiarities of womanhood, its gentle graces, its purity, its emotional impulses, its subordination of hard reason to sympathetic feeling, are not surely qualifications for forensic life. It would be revolting to all female sense of the innocence and sanctity of their sex, shocking to man's reverence for womanhood, that women should be permitted to mix professionally in all the nastiness of the world which finds its way into the courts of justice.'

Re Goodell, Wisconsin Supreme Court, 1875[1]

In 1912, Gwyneth Bebb applied to the Law Society to be registered as what we now call a trainee solicitor. The Solicitors Act 1843 provided that any 'person' with the requisite qualifications, which Gwyneth Bebb had, was entitled to train to be a solicitor. The Law Society, however, refused to allow the application on the grounds that women were not, and should not, be solicitors. Gwyneth Bebb argued that her application was entitled to be accepted as she was a 'person' within the terms of the 1843 Act and she took her fight to the Court of Appeal. However, she lost her case; the judges decided that she was not a 'person'.[2] In the face of all common sense, the judges upheld the Law Society's view, and refused to allow women to become solicitors. It took the suffrage movement and World War One to force the hand of the judiciary; and women were allowed to practise as lawyers after the passage of the Sex Discrimination (Removal) Act in 1919.

Although the 1919 Act did not give rise to the predicted 'dreaded flood'[3] of women solicitors, by the 1960s and 1970s, as a result of the women's movement and the expansion of higher education, women began to enter the solicitors' profession in ever increasing numbers. Now, with the path to progress more open, around 2,000 women each year pursue Gwyneth Bebb's aim and become trainee solicitors. This chapter examines the steps which must be taken to become a solicitor, looking first at the Legal Practice

Course (LPC), and then at the requirement to complete two years (full-time) practical training. In addition, this discussion will shine some light on the extent to which entry into the profession is open to all, free from discrimination. The chapter concludes with the personal testimonies of two women solicitors which, together with those at the end of chapter four, expand on and exemplify many of the themes addressed in this and the following chapter.

INTO TRAINING: THE LEGAL PRACTICE COURSE

The LPC runs for one academic year full-time, or two years part-time, and combines the teaching of substantive areas of law (with both compulsory courses and some options), with the development of skills, including drafting, interviewing, advising and advocacy. The Law Society, as regulator of the course, sets the number of places available at each institution, and in 1997–98, there were just over 6,800 full-time and around 1,300 part-time places.

Applications to study the LPC full-time, at one of the many institutions which offer the course, must be made via the Central Applications Board,[4] with applications for part-time courses being made directly to the institution concerned. No information is available on the respective number of applications and acceptances onto the LPC of women, men and ethnic minority students. Such information would be very useful as it would show who is applying for the course and the likelihood of their application being successful, information which is available in respect of university places.[5]

Law Society research found that although academic performance was the most significant factor in determining the allocation of places on the LPC, this was not the sole criterion.[6] It was discovered that certain types of applicant – those from ethnic minority groups, those who attended a new university and those who had taken a law degree – were significantly less likely than others to have received an offer, even when academic performance was taken into account. This bias was most marked for applicants with lesser academic qualifications. Further, as we will see regarding the allocation of training contracts, there was a distinct bias in favour of students who had taken the Common Professional Examination (CPE), again even when academic performance had been taken into account. Such bias indirectly disadvantages ethnic minority law students and those from less privileged backgrounds as they are more likely to do a law degree than a CPE course.[7]

Given this evidence, suggestive of discriminatory practices and attitudes, the absence of information about the LPC applications process must be remedied. Greater openness would reveal patterns, enable questions to be asked about recruitment practices of particular institutions, allow action to bring about greater equality and fairness and improve accountability. If the profession as a whole is to eradicate discrimination within its ranks, it must begin to take action at the point of entry.

Money, money, money

Embarking on the LPC involves a considerable financial commitment. Tuition fees range from around £4,000 to £6,000, and of course there are living and accommodation expenses on top of this. Joanne Hall has just qualified as a solicitor working in criminal and matrimonial law and had to take out a bank loan for £4,000 for, as she says, the privilege of carrying on her legal work, and this was at a time when she did not have a job to pay off the loan. 'Becoming a lawyer has cost me approximately £7,000', says Joanne, 'which my bank informs me I'll be paying back until 2009. You have to be very determined to want to enter the profession and not give up after your degree.'

Although local authority grants are available, they are few and far between and their number has greatly decreased in recent years. Some firms do offer sponsorship to those who have already accepted a training contract, with amounts varying from firm to firm, ranging from, for example, £500, to the payment of full fees plus a living allowance. The availability of sponsorship depends heavily on the type of firm, with large City commercial firms offering the greatest financial support. The amount of financial support does not mark the importance which a firm attaches to trainees, but reflects the reality that, the smaller the firm, the less the firm can afford. This unfortunately means that public interest, legal aid, and high street firm work demands a financial commitment which commercial work does not.[8]

The introduction of student loans will make this situation much worse. Students, already graduating from university with increasing debts, will be required to take on further liabilities if they wish to pursue a career in the legal profession. In addition, there is evidence that women students earn less in their vacation jobs than men, which, together with the fact that women earn on average less than men in the profession, means that the financial burden may fall heavier on women than on men.[9] If we are committed to an open and diverse legal profession, and do not wish to return to the days when only a select elite could become lawyers, issues of access and funding have to be tackled. In this context, the Trainee Solicitors Group (TSG) has proposed a radical restructuring of the training contract involving a shorter LPC and more training 'on the job'. The aim of the proposals is to ensure that 'merit not means' is the principal factor in choosing trainees and in enabling trainees to succeed.[10] The present review of the training contract being undertaken by the Law Society needs to scrutinise access, given its obvious impact on the future composition of the profession.

Passing the LPC

Seven out of ten students pass the LPC first time.[11] The data available does not break down the statistics by sex or ethnicity. Statistics provided by the Law Society regarding the predecessor of the LPC, the Law Society Final (LSF) exams, were disaggregated according to sex and showed that women had a far higher pass rate than men.[12] As women and those from ethnic minorities appear to have greater difficulty in gaining places on the LPC and training contracts,[13] it is important that full statistics are published regarding the relative success rates of women and men and ethnic minority students. Law Society research shows that women do have a higher pass rate than men on the LPC, and are awarded distinctions in far higher number.[14] This is important information which can feed into an analysis of why women, although achieving on average greater exam success, find it more difficult to gain entry to the profession.

EXPERIENCING WORK

Making decisions about future career choices can often seem inordinately difficult. Narrowing it down to working in the law helps, but only a little. Even reducing it further to the legal profession, or the solicitors' profession, still leaves many options. One of the ways to reduce the dilemmas is to try to gain some work experience.[15] Solicitor Joanne Hall practises in criminal and matrimonial work and first experienced the law when she spent a week with a barrister in Newcastle Crown Court. 'Unfortunately, every case was settled at the doors of the court', says Joanne, 'so I never actually saw a trial take place, but I was hooked on the drama of the courtroom and decided then to try to enter the profession'. Before beginning her training with a small three partner firm in London, freelance solicitor Natalia Siabkin did some vacation work with both a large City firm and a general high street practice. These experiences enabled her to decide that it was in a small firm that she wished to work, with what she calls 'real life legal practice basically dealing with the problematic relationships of individuals, whether in the context of families, employment or housing'. On the other hand, for solicitor Jane Hyndman, work experience in a small high street general practice in central London showed, Jane says, that 'I was not cut out for working in a small general practice as I had no real interest in conveyancing and wished to avoid matrimonial work at all costs'. Jane qualified with a City law firm, and now works in-house for the company Live TV.

What is instructive about these examples is that although the work experience was not in the direct fields in which Joanne, Natalia and Jane have ended up working, the experience helped them decide what it was they wanted to do and to learn more about what it means to be a lawyer. In addition, carrying out some work experience demonstrates to a future employer an interest in and commitment to the law and legal profession.

Furthermore, many firms use work experience placements as a means of recruitment. Although according to Law Society guidelines firms are not technically allowed to interview candidates for training contracts before the beginning of the September in the student's final year, in effect many do in practice by holding 'informal' interviews. Further, recent research found that over one quarter (27 per cent) of those who had gained work experience with a solicitor were offered a training contract by the organisation with which they had gained experience.[16] There is a worrying facet to this trend. The researchers found that it was those who were from ethnic minorities and less prestigious universities who were least likely to get work experience.[17] Thus, a bias at the level of work experience can have a knock-on effect in reducing the chance of securing a training contract.

FROM LEARNING TO EARNING

Getting into training

Having successfully completed the LPC, the next step to qualification as a solicitor is the vocational stage of training, that is two years' (full-time) experience on the job as a trainee solicitor.[18] The aim of a training contract is that the trainee solicitor learns how to put into practice her knowledge of the law. A trainee is supposed to gain experience in at least three areas of law from a list laid down by the Law Society, and the areas of law covered will depend on the work of the firm.[19] Training standards vary from firm to firm and are monitored by the Law Society.[20] Helen Davies, chair of the Law Society's committee reviewing the training contract, has said that there is scope for improving the standards of training contracts. In particular, Ms Davies flagged up the link between poor salaries and poor training, and suggested that the monitoring of training contracts might need to be improved.[21]

Although the vast majority of trainees (96 per cent) do their training in private practice, other training opportunities are growing. Training contracts are available in commerce and industry, in government departments (local and national), in the Crown Prosecution Service and advice services. What is interesting about these opportunities is that many more women than men gain training contracts in-house.[22] This is a pattern which is replicated for solicitors, with proportionately more women than men practising as a solicitor in-house.[23] The reasons why it might be that more women than men practice in-house will be considered in the following chapter.[24]

Who are trainee solicitors?

Table 3.1 shows that in 1996–97, 53 per cent of traineeships registered were for women, and there was a total increase in the number of trainees from the previous year of 17 per cent. Since 1987–88 there have been more women trainees registered than men, reaching a peak in 1991–92 when 55 per cent were women. Since then, the representation of women has been falling. This has happened at the same time as the numbers of women law students has continued to increase. However, in view of the lack of information, noted above, it is not clear whether these figures reflect lower numbers of women taking the LPC, as such statistics are not published, though they should be easily obtainable. It will be very interesting to watch how the numbers of women progress over the next few years to see if a pattern is discernable. Will it be that, despite the increasing number of women law students, together with the fact that women have, on average, far better academic credentials than men, equal numbers of women and men are being recruited into the profession?

Table 3.1: Traineeships registered 1985–86 to 1996–97 by sex of trainee[25]

Year	Men	%	Women	%	Total
1985–86	1,581	54	1,337	46	2,918
1986–87			(Figures not recorded)		
1987–88	1,407	48	1,511	52	2,918
1988–89	1,464	48	1,594	52	3,058
1989–90	1,559	48	1,695	52	3,254
1990–91	1,771	46	2,070	54	3,841
1991–92	1,783	45	2,158	55	3,941
1992-93	1,677	46	2,004	54	3,681
1993–94	1,815	47	2,050	53	3,874
1994–95			(gender not recorded)		4,170
1995–96			(gender not recorded)		4,063
1996–97	2,230	47	2,509	53	4,739

The geographical spread of training contracts is shown in table 3.2 which shows that just over one quarter of all training contracts are registered in the City of London, and that a further fifth are registered in the rest of London. Overall, therefore, London is the home of 47 per cent of trainees. Male trainees are more likely to be in London than women (49 per cent of men and 47 per cent of women), and the area with the highest concentration of women trainees is the North (61 per cent).

Table 3.2: Traineeships in 1996–97 by region and sex[26]

Region	Men	Women	Total	% women
City of London	629	637	1,266	50
Rest of London	455	506	961	53
Rest of S East	212	287	499	58
South West	147	161	308	52
Wales	80	107	187	57
West Mids	172	156	328	48
North West	217	264	481	55
North	43	68	111	61
Yorkshire and Humberside	143	135	278	49
East Mids	106	154	260	59
East Anglia	26	34	60	57
Total	2,230	2,509	4,739	53

It can be seen from table 3.3 that in 1996–97 just over one in ten trainees were from ethnic minorities. Overall, the proportion of women was higher among ethnic minority trainees; 53 per cent of all trainees are women, but 57 per cent of all ethnic minority trainees are women. Comparative figures for previous years are not available, although we know that ethnic minorities make up 13 per cent of admissions to the profession, a figure which has remained largely constant since the early 1990s.[27]

Table 3.3: Number of trainees in 1996–97 by ethnic origin[28]

Ethnic Origin	Men	Women	Total	% women
Afro-Caribbean	14	18	32	56%
Asian	158	179	337	53%
Chinese	23	45	68	66%
African	12	22	34	65%
White European	1,850	2,068	3,918	53%
Other ethnic origin	35	55	90	61%
unknown	138	122	260	
Total	2,230	2,509	4,739	53%
Total ethnic minority	242 (11%)	319 (13%)	561 (12%)	57%

Law Society research analysed the success factors regarding the allocation of training contracts and found that, as with applications to the LPC, although academic performance was a crucial factor, it was not the only consideration. There was a clear bias in favour of CPE students and those who had completed vacation work placements.[29] The other factors which affected outcomes, regardless of academic performance were sex, ethnicity, social class background and type of institution attended (new or old university), although sex and the type of school attended mattered less than the other factors. When considering CPE graduates alone, sex and ethnicity made no difference. The researchers concluded that 'in the allocation of training contracts, the solicitors' profession appears to have a definite bias towards white males who were educated at an independent school, Oxbridge or the College of Law'.[30]

This evidence reveals further worrying trends regarding access to the solicitors' profession. It is not clear why CPE students are considerably more likely to gain a training contract, although chapter one examined some of the possible reasons for this.[31] In addition, the tendency of training contracts to be granted to those of a particular social class and educational background suggest that they are not being awarded on the basis of credentials alone, and other factors, often attributes largely beyond an individual's control, are important.

FINDING A PLACE TO PRACTICE

It can often seem very difficult to choose between different firms when seeking a training contract. The area of work and place of work will be important, as will the number of trainees that a firm keeps on at the end of the training period. The variety of the work on offer and how seriously training is taken will also be significant. The remainder of this section will consider two further factors which may be borne in mind; the salary and the firm's attitude towards equal opportunities.

Paying the minimum

At one time, a trainee solicitor, then known as an articled clerk, had to pay for the privilege of being trained. Although trainees are now paid during their training, the amount of the salary varies tremendously. Trainees in the larger City firms will often be starting on a salary of around £17,000 to £20,000, whereas the reality for many is a salary about half that size, with many paid even less. In a recent survey of trainee solicitors carried out by the TSG, the highest paid trainee was getting £23,000, and the lowest just £6,510 per annum.[32] In order to try to maintain the semblance of a living wage for trainees, the Law Society enforces a minimum salary for all trainees (with exceptions possible), the present rate of which is £10,850, rising to £12,150 in central London.

There have been periodic moves to try to remove the minimum salary for trainees, and its existence is under review again as part of the general review of training contracts referred to above. Those in favour of abolition tend to suggest that this would create more training contracts. However, there is little evidence to demonstrate this, and indeed, there is much evidence to suggest a strong relationship between low wages and poor training. Moreover, there is a further dimension to this debate which must be taken into account. The research carried out by the TSG also demonstrated that women trainees were paid on average £779 less per annum than men, increasing to a difference of £1,192 in legal aid/high street firms. Further, for every two men on the minimum salary, there were three women, and 50 per cent of ethnic minority trainees were on the minimum, compared with only 26 per cent of trainees as a whole. Although the sample in the TSG study was small, it is suggestive of lower wages for women, findings which are also supported by the Law Society's figures.[33] Table 3.4 shows that salaries for men are two per cent higher than for women. The Law Society does not provide any statistics regarding the comparative salaries of ethnic minority and other trainees.

It appears possible, therefore, that the abolition of the minimum salary may adversely affect women and ethnic minority trainees; those who are already at a disadvantage when it comes to applying for the LPC and training contracts. The impact of this evidence goes beyond the minimum salary debate. The next chapter considers the fact that women solicitors and partners are paid substantially less than their male counterparts.[34] The TSG research and the Law Society's figures are suggestive that the problems begin at the level of training. Thus, the common assumption that because there are equal numbers of women and men gaining training contracts, the problems of discrimination only arise later, does not stand up to scrutiny. We must look beyond the bare statistics and see that there are many forms of discrimination against women and ethnic minorities.[35]

Furthermore, we have already established that women on average gain superior academic credentials to men. The fact that men on average are being paid more as trainees suggests that academic credentials may not be the principal factor in the allocation of training contracts. Men appear to have greater 'purchase' than women when it comes to setting their remuneration. Is this because, from the beginning, law firms do not see women as a long term investment? Do they make certain gendered assumptions about the capabilities and career patterns of women? Do they just value women less? These are difficult questions, but ones which must be faced.

Table 3.4: Trainees' starting salaries by region and sex in 1996–97[36]

Region	Male average	Female average	All trainees average	% male average over female	% trainees at or below min salary
Central London	17,409	17,302	17,355	0.6	11
Rest of London	14,223	13,532	13,858	5.1	24
Rest of S East	11,668	11,704	11,689	- 0.3	46
South West	11,723	11,722	11,723	-	39
Wales	10,671	11,039	10,882	- 3.3	66
West Mids	11,599	11,885	11,733	- 2.4	48
North West	11,715	11,691	11,702	0.2	55
North	11,115	11,250	11,199	- 1.2	58
Yorkshire and Humberside	11,938	12,106	12,019	- 1.4	45
East Mids	11,411	11,475	11,449	- 0.6	52
East Anglia	11,610	11,560	11,582	0.4	49
Total	**14,037**	**13,757**	**13,889**	**2.0**	**33**

Representation of women solicitors and partners

Another factor to consider is the gender make-up of the firm or organisation to which an application is being made. The mere fact that, for example, a firm employs equal numbers of women and men trainees does not per se indicate that a firm is committed to equality, but it does at least suggest a certain level of awareness and equity. In smaller firms, the question may be not so much equality in numbers at the time of commencing a training contract, as a firm may only take on one or two trainees, but the firm's track record over the past, say, five to ten years.

The more interesting statistics are likely to be those regarding the number of women partners, and particularly where there are few partners, the attitude of a firm to such low numbers of women partners. As we will see in more detail in the next chapter, there are not only fewer women than men partners as a whole, but there are also fewer women being made partner.[37] Law Society statistics show that of those solicitors in private practice with 10 to 19 years' experience – with sufficient experience for partnership – 88 per cent of men are partners or sole practitioners, compared with only 63 per cent of women.[38] Thus, there are women solicitors in firms, with the requisite experience, who are simply not being made partner at the same rate as men.

In 1995, the group Young Women Lawyers (YWL) carried out a survey of the largest 100 solicitors' firms which sought information on the numbers of women being made partner in that year. The survey found that only 25 per cent of new partners were women.[39] Although the vast majority of law firms did not consider this to be a problem, no excuse can be made for these statistics. Women and men have been entering the solicitors' profession in equal numbers for around ten years. This means that unless the numbers of women being made partner improve drastically, the low numbers of women will continue. Thus, any firm suggesting that it is only a matter of time before women are making partner at the same rate as men simply does not understand the extent or nature of the problem.

Caroline Graham recently changed jobs, moving from a large Newcastle firm, to the City. In her interviews with various firms she asked generally about the 'glass ceiling' and the number of women partners. If women did not seem to be breaking through, she asked interviewers if they had any views about why this might be. 'The firm which explained that its women solicitors leave in greater numbers than men received an immediate rejection of their job offer', says Caroline. 'This is not an answer, but begs many questions which this firm had not even begun to address. I am joining the firm which said that women were receiving a tough deal within the profession and that, whilst it may be difficult at times, firms had to work to create the kind of flexibility which will enable women to succeed.' Caroline urges women in the job market to raise such questions at interview, even though she recognises that this is a 'counsel of perfection, especially when competition for training contracts is so intense'.

Closely allied to the above discussion regarding the numbers of women in a firm is the question of the employment conditions of each firm or organisation. In many ways, it is the culture, attitudes and employment practices of a firm which can greatly affect the number of women in the organisation. We will see in the next chapter that this may be one of the reasons why women are increasingly choosing to work as solicitors in-house.[40] Many companies and other organisations are more advanced in their thinking regarding employment conditions than many solicitors' firms, although it is difficult to generalise. For example, a firm's policies regarding maternity, paternity, parental leave, equal opportunity and equal pay send out clear signals regarding that firm or organisation's attitude to the employment of women. The need for such policies is considered in the next chapter.[41]

These factors are important to women entering the profession and firms should be more aware that these issues will influence decisions as to where to practice. Research carried out in 1997 by YWL found that nine out of ten women law students said that equal opportunities policies in the firm would be important in determining where to train.[42] Three quarters said that flexible working hours were important; three out of five emphasised maternity pay above the statutory minimum and child care facilities or financial help therewith, and just under half said that the facility to work from home would be important. Although it is sometimes difficult to find out whether firms or chambers have such policies, it is important to do this kind of leg-work while on work experience or at law fairs. In addition, it will only be when such questions are routinely asked, that firms and chambers will begin to recognise that they must change.

CONCLUSIONS

This chapter has examined the steps which must be taken in order to qualify as a solicitor, and we have seen that, although this requires determination as the competition is often stiff, the intellectual and personal rewards can be very high. As the personal testimonies at the end of this and the following chapter show, being a solicitor can open up a vast range of opportunities, from working in a large City firm, to being a sole practitioner, to working in central or local government, to being part of the legal team of a company or building society, to working in a law centre, to name just a few possibilities[43]: being a solicitor is indeed, as solicitor and chair of the Equal Opportunities Commission Kamlesh Bahl suggests, an 'excellent option'. Accordingly, solicitor Joanne Hall says that she 'would encourage women students to enter what is an exciting and challenging profession'.

This chapter has also considered questions of access to the solicitors' profession. The legal bar which Gwyneth Bebb faced in 1912 was long ago removed, but it appears to have been replaced by informal and more subtle barriers, affecting ethnic minority students in particular. The profession must ask itself some difficult questions regarding these issues, and institute reforms: open access, free from discrimination, is essential for the future composition and culture of the profession.

PERSONAL TESTIMONIES OF WOMEN SOLICITORS

Kamlesh Bahl CBE, Chairwoman of the Equal Opportunities Commission, Deputy Vice-President of the Law Society, 1998–99

Kamlesh made around 250 applications for a training contract and was made only one offer. She took the job with the Greater London Council and qualified as a solicitor. She then went on to work as an in-house solicitor for a number of companies. Since 1993 she has been chairwoman of the Equal Opportunities Commission which involves her in providing strategic leadership and in being the external spokesperson for the organisation. Kamlesh relates below some of her experiences and views on women in the law ...

'I have four sisters and one brother and I am the only lawyer in my family. My choice of a legal career was inspired very much by the influence which my uncle had on my mother when she was growing up in Kenya. My uncle trained as a barrister in this country and went back to practice in Kenya and used to take my mother to his office and the courts with him on a regular basis. He had a very successful career and was ultimately appointed as Chief Justice.

My first experience of discrimination was when I was looking for articles in my final year at the University of Birmingham. Despite having come first in my first year and being in the top three in the second year, I found it extremely difficult to get a job. The careers advice given to me was that I should aim for jobs in the big City firms because of my excellent qualifications. This experience taught me two things. First that discrimination is a tremendous waste of resources. Here I was, well qualified, eager and willing to start my career and simply wanting an opportunity to do so, which was being denied to me. Secondly, that the tried and tested path might not be appropriate for you. Either as a woman or as a member of an ethnic minority, I knew with certainty that I could get to where I wanted to but I would just have to find my own route, and not use the ones which had been tried before.

It is important for women in the law to recognise that there is discrimination, but not to be disheartened by it. There are many ways to get to the same goal and you may need to find or create your own path. The increasing number of women in the legal profession has helped to change attitudes within it and is helping to change working practices so that a career as a lawyer is an excellent option for women. The future prospects for women in law are very good indeed. The legal profession, as in any other business, needs to recruit the best talent and ability for it to be successful. This is increasingly recognised and valued, and will have a beneficial impact in terms of the employment of women.'

Rachel Tetzlaff-Sarfas, partner, Thompsons

Rachel spent a number of years trying out various jobs and opportunities before deciding that being a plaintiff personal injury solicitor was the job for her. She is now a partner in the firm Thompsons and describes below her experiences and views on the future for women in the law ...

'I have been fascinated with the law more or less since my early teens when I first began to read sensational reports of criminal trials in my parents' newspaper. Justice, fairness and equality have always been important principles to me. I had a social conscience and I saw the law as the tool I could use to uphold my beliefs. I wanted not only to be a champion of justice but also to make the law accessible.

The careers' teacher at my school thought I was being over ambitious when I announced that I was going to be a lawyer. Indeed, she did try hard to persuade me that I was not quite up to it! I did not do brilliantly at my A levels, but I did pass. I was resolute. I even arranged to work at a local law firm one school afternoon a week to find out just what I was letting myself in for. In spite of my strong conviction that I wanted to be a lawyer, I did not feel inspired by the idea of studying law. I thought that it would be dull.

I was persuaded that, in view of my chosen career, any other subject would simply be a waste of time. Although I was never entirely sure quite what the rush was about, I eventually compromised and went off to Warwick University to study Law and Sociology, straight from school.I ended up at Warwick more than through luck than judgement, although it was ideal for me. The course placed the law in a social context and was not all the dry textbook learning that I had worried about. I graduated in 1988 but still did not feel ready to embark on my long chosen career. Whilst at university, I had toyed with the idea of becoming a Probation Officer and had worked once a week in a local probation drop-in-centre for two years. On leaving university, I gave up the coveted place that I had reserved at Guildford College of Law and went off in search of further work experience. I worked as a Housing Assistant and an Employment Officer before doing a stint as an environmental campaigner in California and as a bank clerk in Sydney, Australia.

Finally, two years after graduation, when I had travelled around the world and back again I decided it was time to become a solicitor. The only problem by then was that half the other graduates in Britain had a similar idea. I was unable to secure a place on the then equivalent to the LPC. I persevered. I wrote about forty speculative applications to firms specialising in legal aid. Miraculously within three weeks I had two job offers. I started work as an outdoor clerk and soon signed a contract for articles. I sandwiched my two years of articles either side of the Law Society Finals course at Store Street and managed to pass the course with honours.

I took a rather circuitous route to my chosen career. And yet when I finally began my training, I was still not exactly sure what kind of law I wished to practice. I knew that it would be easy to fall into a discipline that did not inspire or thrill me and then to be stuck in it until retirement (I had certainly met enough lawyers like that during my training). Luckily I spent my final few months of articles in the Plaintiff Personal Injury department. It only took me a week or so to decide that I wanted to practise personal injury law. For me this was where law touched 'real' people's lives. As a plaintiff personal injury lawyer I could battle against insurance giants and satisfy my social conscience. I was offered a job at the firm where I did my articles but it was not in the right department. I knew it was a gamble but I desperately did not want to be stuck in a job that I did not enjoy. I took out the Chambers directory and trawled through applying speculatively to Plaintiff PI firms. Bingo! I struck lucky again and my first interview landed me the job of my dreams with Robin Thompson and partners as a PI lawyer. I am still with the same firm which is now known as Thompsons. It is the largest plaintiff personal injury firm in the country. Many of my clients are trade union members and I can now pride myself on only ever acting for the injured party. After just over three years with Thompsons, I was made a partner. My gamble paid off and I am fortunate enough still to get a kick out of my job.

If you choose to study law at 18, as I did, it is very easy to be swept along on a lawyer's band wagon. I was encouraged to apply to law school and for articles almost before my first university term was complete. I consciously chose not to take that route. I wanted to be sure that I could find a job that I could do something that I believe in. Thompsons is a young, dynamic firm where I can be myself.

I believe that I have experienced discrimination whilst at work. Many years ago I discovered that a male colleague doing the same job as I, had been earning £1,000 pa more. I am ashamed to say that I swallowed it at the time but have felt sick about it ever since. I was afraid of the consequences of speaking out. I worked in an all male environment and had no ally to turn to. Support networks are essential and need to be accessible. I do not know what I would have done had someone been there to support me. As a woman I want to be treated as an equal and not with favouritism. Open salary scales might not eliminate inequality in pay between men and women but it would certainly give women the information necessary to begin to effect change. After all knowledge is power and women certainly need a lot more of that for further progress to be made for women in law.'

NOTES

1   *Re Goodell*, 39 Wisc 232,1875, p 245.
2   *Bebb v Law Society*, [1914] 1 Ch 286.
3   Harry Kirk, *Portrait of a Profession*, Oyez: London, 1976, p 112.
4   Contact details can be found in appendix five.
5   See chapter one under the heading 'Who succeeds? law school applications and acceptances' and table 1.1.
6   Michael Shiner and Tim Newburn, *Entry into the Legal Professions – The Law Student Cohort Study Year 3*, Law Society: London, 1995, pp 42–58 and Michael Shiner, *Entry into the Legal Professions – the Law Student Cohort Study Year 4*, Law Society: London, 1997, p 25.
7   See chapter one under the heading 'Common Professional Examination' and Michael Shiner, ibid.
8   Loans are available from many banks, and the Department of Education and Employment offer what are called Career Development Loans (see *Support for Students – a guide to grants, loans and fees in higher education 1998–99*, free from the Department for Education and Employment on (tel) 0800 731 9133). A Law Society Bursary Scheme has limited awards for the LPC, CPE and Postgraduate Diploma in Law which are granted on the basis of competitive elements and hardship criteria. Some universities and publicly funded colleges may provide access funds to students, likely based on hardship criteria, and the Law Society processes a number of limited scholarships which are available to students from ethnic minorities.
9   One study found that women students earned on average £3.80 an hour, with men earning £4.55: *Times Higher Educational Supplement*, 25 July 1997.

10   Nick Armstrong, 'The training contract under review', 148 (1998) *New Law Journal* 216–218.
11   Bill Cole, *Trends in the Solicitors' Profession – Annual Statistical Report 1997*, Law Society: London, 1997, p 67.
12   Law Society, *Trends in the Solicitors' Profession – Annual Statistical Report 1994*, Law Society: London, 1995, p 68.
13   Michael Shiner, supra note 6, pp 25, 67.
14   Michael Shiner and Tim Newburn, supra note 6, p 68.
15   The Law Society and Higher Education Careers Service have recently established a list of all vacation placements within law firms. The list is available at: http://www.prospects.csu.ac.uk.
16   Michael Shiner and Tim Newburn, supra note 6, p vi.
17   Ibid p 41.
18   For more information on training contracts, see *The Training Contract Handbook 1998* produced by the TSG and available through the Law Society. For information on the TSG, see appendices five and six.
19   During the training contract, a trainee must also complete the Professional Skills Course which covers accounts, investment business, personal work management, professional conduct and advocacy. The course lasts for up to twenty days. Its existence, however, is presently being reviewed by the Law Society because of opposition to it from many law firms.
20   To whom a complaint about the training received can be made. The TSG may also be a useful point of contact in respect of any problems, and in particular they run a number of helplines for those experiencing problems with their training.
21   *Law Society Gazette*, 23 July 1997.
22   63 per cent of non-private practice trainees in 1996–97 were women: supra note 11, 73.
23   Supra note 11, p 15: 84 per cent of men work in private practice compared with 76 per cent of women.
24   See chapter four under the heading 'In-house and sole practice: is the grass greener?'.
25   Supra note 12, p 65 and note 11, p 72.
26   Supra note 11, p 73.
27   Ibid p 83.
28   Ibid p 72. The total for ethnic minority trainees excludes those whose ethnic origin is unknown.
29   Michael Shiner and Tim Newburn, supra note 6, pp 84–5.
30   Michael Shiner, supra note 6, p 52.
31   See chapter one under the heading 'Common Professional Examination'.
32   Nick Armstrong and Richard Moorhead, 'Bare minimum', 147 (1997) *New Law Journal* 487–489.
33   Michael Shiner, supra note 6, p 92.
34   See chapter four under the heading 'Paying for equality'.
35   See also Catherine Barton and Catherine Farrelly, 'Women in the legal profession – a student perspective', 148 (1998) *New Law Journal* 599.
36   Supra note 11, p 74.
37   See chapter four under the heading 'The sisyphus factor'.

38   Supra note 11, p 17.
39   Clare McGlynn and Caroline Graham, *Soliciting Equality: Equality and Opportunity in the Solicitors' Profession*, Young Women Lawyers: London, 1995. For further information on YWL, see appendices five and six.
40   See chapter four under the heading 'In-house and sole practice: is the grass greener?'.
41   See chapter four under the heading 'Towards Equality'.
42   YWL Press Release, 'Women Law Students Anticipate Discrimination', 3 February 1998, and quoted in YWL Newsletter, January 1998.
43   See further Fiona Boyle, *Legal Careers*, Fourth Estate and Guardian Careers Guides, 1997.

Chapter 4

Soliciting equality: women in the solicitors' profession

'I do not believe that the active practice of a profession is compatible with the proper work of a woman which, after all, is that of a wife and a mother attending to her family. I recognise, of course, that there are a great many women who ... have not the opportunity of marrying. ... Possibly, in the course of time, means may be evolved of enabling them to find a sphere in other parts of the British Empire where they may become mothers of mighty nations.'

The Lord Chancellor, Lord Finlay, 1917[1]

'It is difficult to see why up to now there have been ... no female lawyers in the UK. Clearly, both branches of the law offer as excellent an opening [for] a celibate woman with exceptional talent than as any other profession ... There can be no doubt that at least one per cent of women are quite as intelligent as any man.'

Concerning Solicitors, 1920[2]

The entry of women into the solicitors' profession was resisted long after the legal barriers had been removed with the 1919 Sex Discrimination Act. Central to this opposition was the dominance of an ideology of the 'separate spheres' of the sexes; in other words, that women were suitably confined to the private sphere of home and hearth, with men as the rightful occupiers of the public sphere of work and politics. This division was thought appropriate as it was generally considered that women did not have the intellectual capacity for the law, or any other public duties, as the second quotation above demonstrates. It was also thought that those women who did exercise their brains were putting their very health at risk, and in particular their fertility: the so-called 'womb-brain' dilemma facing women.[3] Indeed, it was considered that menstruation induced a period of 'temporary insanity' in all women.[4] Women lawyers, therefore, faced a number of informal barriers to their advancement, obstacles premised on the notion of women as emotional, at the mercy of their hormones, lacking in intellectual capacity and far better suited to the home than work.

Attitudes have undoubtedly changed and it bodes well for the future that women now account for over half of those admitted to the profession each year and one third of all practising solicitors.[5] Nonetheless, although more women are entering the profession in ever increasing numbers, and some are undoubtedly achieving great successes, barriers still face women when it comes to reaching senior positions, and to being treated as equals in the working environment. After examining the present status of women solicitors, and the ways in which many women remain disadvantaged in the profession, this chapter considers some of the means by which progress and change can be encouraged. The chapter concludes with the personal testimonies of three women working in the solicitors' profession which, together with those at the end of chapter three, exemplify and expand on many of the themes identified in this chapter.

PRESUMED EQUAL

The sisyphus factor

We can see from table 4.1 that, since 1983, women have comprised over one third of those admitted to the profession, and over one half since 1992–93. The proportion of women compared to men being admitted peaked in 1994–95, but has actually fallen back since then. Further, although equal numbers of women and men are admitted to the profession, different patterns of employment begin to emerge soon after admission.

After admission, not everyone goes on to practice, but those who do are required to hold a practising certificate (PC), issued by the Law Society. Far fewer women than men hold a PC.[6] This does not necessarily mean that more women than men have left the profession, but it may be that they are working 'in-house' in institutions for which a PC is not required. Figures relating to solicitors' categories of employment show that 84 per cent of men work in private practice, compared with only 76 per cent of women. Thus, in private practice, women make-up three in ten solicitors; whereas in all non-private practice areas, they comprise four in ten solicitors. There are some indications that women find a more hospitable working environment outside private practice, and these issues are explored towards the end of this chapter.

Table 4.1: Men and women admitted to the solicitors' profession 1923 to 1996–97[7]

Year	Men No.	%	Women No.	%	Total
1923	436	98.2	8	1.8	444
1948	857	97.7	20	2.3	877
1963	768	95	37	5	805
1973	1,542	87	222	13	1,764
1983	1,637	63	959	37	2,596
1988	1,750	54	1,494	46	3,244
1988–89	1,834	53	1,600	47	3,434
1989–90	1,990	53	1,739	47	3,729
1990–91	2,238	52.5	2,027	47.5	4,265
1991–92	2,280	51	2,184	49	4,464
1992–93	2,160	49	2,257	51	4,417
1993–94	2,281	47.5	2,520	52.5	4,801
1994–95	2,229	47.5	2,466	52.5	4,695
1995–96	2,203	48	2,417	52	4,620
1996–97	2,590	48	2,827	52	5,417

Remaining within private practice, statistics show that the higher up firm hierarchies one looks, the fewer women one sees. Once qualified, solicitors will usually be referred to as 'assistant solicitors', a term which includes solicitors with a wide range of experience. Therefore, although we know that in 1996–97 there were equal numbers of women and men assistant solicitors, we do not know the number of senior women assistant solicitors. Increasingly, however, firms are introducing the designation 'associate solicitor', connoting a degree of seniority between assistant solicitors and partners. In 1996–97, 45 per cent of associate solicitors were women. Nonetheless, it is important not to read too much into such a figure as only just over 1,500 solicitors are so designated. A 1995 survey by the group Young Women Lawyers (YWL) found that of those assistant solicitors with zero to three year's experience, 50 per cent were women, falling only to 40 per cent of those with more than three year's experience.[8] This is suggestive that there may be more senior women solicitors than is often thought to be the case, although this data still does not tell us how many women there are at, say, six to ten year's experience when partnership decisions may be being made.

Women partners accounted for only 16 per cent of all partners in 1997, a figure which has increased by only three per cent since 1993.[9] To a limited extent this figure is historically understandable, however time alone does not account for the absence of women partners. For example, the 1995 YWL survey found that of those partners being made up that year, only 25 per cent were women.[10] Moreover, the overwhelming majority of solicitors' firms said that they did not consider there to be a shortage of women eligible for partnership: they were happy with the figure of 25 per cent. A follow-up survey in 1997 found that 27 per cent of new partners in the largest 100 firms were women.[11] Thus, although there has been some welcome improvement, progress is slow.

Furthermore, many of the firms justified this low figure of new women partners by suggesting that, in time, more women would reach senior levels, as women had not been entering the profession for very many years: the 'trickle-up' theory.[12] However, what these figures demonstrate is that, in time, women might just make-up one quarter of partnerships, but no more. Moreover, this is not a situation which is going to be solved just by time. Women have been entering the profession in almost equal numbers for a sufficient number of years to be making more of an impact at partnership level. The 'trickle-up' argument is also defeated by Law Society statistics covering all firms which show that of those solicitors with 10 to 19 years of experience, 88 per cent of men are partners or sole practitioners, compared with only 63 per cent of women. What this demonstrates is that of those men and women with the requisite year of experience to be eligible for partnership, a man is much more likely to become a partner than a woman. Accordingly, it must be recognised that the mere passage of time will not ensure equality for women in the solicitors' profession and this myth must be exposed.

Ethnic minority women solicitors

In 1996–97, just over one in ten solicitors admitted were from ethnic minorities, a figure which has remained largely static through the 1990s.[13] Women constitute over half (54 per cent) of the ethnic minority solicitors admitted in 1997; comparable figures for previous years are not available. Ethnic minority solicitors represent 4.5 per cent of those with practising certificates, an increase from 4.1 per cent the previous year. In detailing these statistics, the Law Society states that of the economically active population in England and Wales, 5.2 per cent are from ethnic minorities. However, no separate figure is given in respect of ethnic minority women.

The patchy nature of the information available on ethnic minority solicitors, especially ethnic minority women solicitors, makes detailed analysis difficult.[14] For example, there is no data on the partnership rates of ethnic minority solicitors, or on levels of pay. All that we can say is that there appears to be a small increase in the numbers of ethnic minority solicitors admitted each year, but as to how they fare in the profession, the question is open. In view of the evidence of discrimination at the point of entry to the profession, and of racial discrimination in society at large, it cannot be assumed that equality prevails, and therefore further data and research is required.

Paying for equality

Inequality of pay between women and men is also a serious concern in the profession. Any illusion to the contrary was shattered by the publication in 1996 of Law Society research which found that, comparing median earnings, male assistant solicitors earned £24,000, while women earned only £21,000.[15] Even where factors such as size and location of firm, and age and length of qualification were taken into account, the difference in earnings at assistant level was £1,700. At salaried partner level men earned £37,000 compared with £32,000 for women, and at equity partner level the disparity between women and men rose to £15,000. This picture had changed little by the time of publication of figures in 1997, and indeed the gap increased in respect of the pay of assistant solicitors (to £2,340 for assistant solicitors, even taking into account the size of firm, area of work, years of experience and other such factors).[16] The facts are clear: from assistant solicitor to salaried partner to equity partner, women are paid less than men.

The Law Society: governing women?

The solicitors' profession is governed by the Law Society and it therefore falls to be asked what part the Law Society plays in sustaining or alleviating the disadvantaged position of women solicitors. The Law Society's Council comprises 75 members who are either elected from local constituencies or co-opted to represent certain groups of solicitors. There are around ten women council members, which is not representative of women in the profession. As a result, there has been considerable debate over the years about designating one of the seats on the council for a representative from the Association of Women Solicitors (AWS).[17] Thus far, it has only been agreed that one council member should be nominated as a member responsible for raising issues of concern to women.[18] It is important, however, that the person so nominated is representative, or has links with a body such as the AWS, in order that their interventions are based on evidence and actual knowledge of the concerns of women solicitors. Clearly, the best person for this role is a representative from the AWS.

It would also be advantageous for there to be more women elected to the Council. Eileen Pembridge, a council member and the first woman to stand for election as President of the Law Society in 1995, suggests that one of the reasons why women might be put off getting involved is the 'overwhelmingly male' culture of local law societies which are often little more than 'dining clubs'.[19] Council member Lucy Winskell has commented that the first thing that a woman notices about the regular council meetings is that she is in a distinct minority.[20] Affirmative action, Eileen Pembridge believes, is necessary because until there is a critical mass of women on the

Council issues like 'part-time work, the glass ceiling [and] maternity leave for partners' will not be brought to the fore.[21]

In some ways more important than the Council, in terms of its ambassadorial role, is the President of the Law Society: a position which, since 1995, has been elected by a vote of Law Society members. The first elected president, Martin Mears, suggested that those who were seeking to eliminate discrimination within the profession were 'feminist zealots [who] thrive on grievances and use minorities as raw material for their whinge factories'.[22] Martin Mears was not re-elected and subsequent presidents have learned to be more proactive regarding equal opportunities. Kamlesh Bahl, chair of the Equal Opportunities Commission (EOC), solicitor and member of the Law Society Council, has been elected deputy vice-president for 1998–99. This puts her on course to become the first woman and ethnic minority President in 1999–2000.[23] Her election sends a welcome message to the profession, and may lead to a more proactive stance being taken regarding discrimination within the profession.

BARRIERS AND CONSTRAINTS

Explaining it away

'There are no barriers to the advancement of women apart from those they choose themselves or those that are inherent in their biology'.

Martin Mears, President of the Law Society, 1995–96[24]

So far, this chapter has documented the extent of the pay inequality between women and men, and the under-representation of women at senior levels, in the profession. It has also been shown that this situation cannot be justified by reference to historical exclusion of women from public life. Why is it, therefore, that despite equal numbers of women entering the profession for many years, they remain marginalised? In response to such a question, it is often suggested that the under-representation of women is the result of the combined products of the market and women's own choices. This 'human capital' theory suggests that women voluntarily invest less in their education, training and career and it is therefore not surprising that they are less well paid and promoted less often than men who have invested in themselves.[25]

In other words, the 'human capital' women build up is not as great as that of men. Thus, it is often repeated that women who leave solicitors' firms in not inconsiderable numbers are doing so because of their own choice not to invest in their careers/human capital. The conclusion drawn is that it is not the fault of the firm that a woman solicitor leaves because, for example, her request to work part-time is refused or because of the firm's negative attitude towards her second pregnancy. This is seen as the woman's choice to leave, despite the fact that she may have invested years in her career and may not have wished to leave the firm. Although in the end it was she who handed in her notice and walked out the office door, her 'choices' were constrained by the actions and attitudes of the firm, to the extent that, in effect, she was left with no choice.

The human capital rationale also suggests that having a child is a 'choice' that each woman makes, and that this choice necessarily reduces the woman's commitment and investment in her career. Thus, the fact that it is after childbirth that women begin to experience limits on their progress is supposedly justified on the basis that they themselves have signaled to the world, by their ability to procreate, that they are no longer as committed to the workplace as men.

It might seem, thus far, that women who have similar employment patterns to men (that is no family responsibilities, long hours, no breaks in employment) can achieve similar status and pay levels as men, because they have exhibited the necessary 'human capital'. However, even for such women, the human capital theory has a variant which may stand in their way. It is often assumed that intrinsic biological differences between women and men give women a comparative advantage to working in the home and childcaring, and that their decisions not to invest in human capital (themselves) are based on this biological 'advantage' over men.[26] A more extreme version has suggested that 'biology *exonerates* men (and "society") by explaining male economic success in terms of innate, uncoerced differences in ability and desire'.[27]

This line of reasoning suggests that even the woman who does not have children may be hampered by the very association of women with children, the home, their biology, emotions and supportive roles, whereas men are associated with work, authority and rationality. This may affect the manner in which women are treated, for example simply not being considered as authoritative as men. In addition, it is assumed that women will have children, a presumption which affects how employers, and other employees, treat women from day one of their employment. Therefore, when former Law Society President Martin Mears suggests that one of the two limitations on women's careers is biology, this is in fact an overarching limitation which affects all women. Thus, even the woman who tries to become an honorary man is still disadvantaged as ultimately she remains a woman.

An alternative explanation suggests that women's lower pay and status are the combined products of inefficiency, constraint and discrimination, and that gender plays a crucial role in determining women's careers. Thus, the argument in response to the human capital theory is that the profession inefficiently penalises the comparable efforts of committed women who, often in spite of family responsibilities, invest heavily in their careers.

Hence, when considering the number of women who leave private practice, it might be noted that those women have often invested heavily in their careers, which is not suggestive of a lack of commitment. It is simply that such investments are not rewarded to the same extent as men's. One reason why this may be the case is that the masculine culture of the profession has created assumptions about ways of work and images of 'ideal' lawyers to which women may aspire, but most will rarely, if ever, meet. These assumptions are deeply embedded and are central to our understandings of women and men generally. For example, as noted above, women are often associated with specifically feminine characteristics which are not those associated with the authoritative lawyer.[28] The image of authority is masculine, the woman in authority is therefore other than the norm, the exception.

Equally, there is an assumed standard career pattern, premised on the worker having a female partner to take care of family and personal responsibilities. A further aspect of this masculine culture is that those who are other than the norm are not imbued with the same 'cultural' capital. As will be considered, cultural capital concerns the social and personal attributes which have the effect of trumping credentials, such as academic success and experience gained, that is human capital.[29] These explanations will be considered in the following sections by highlighting certain facets of the disadvantages facing many women.

A baby bar?

Whereas once discrimination affected women on their marriage – the marriage bar – it is now clear that it is upon becoming pregnant that women encounter discrimination in the workplace. In one recent case, for example, a woman legal executive won her sex discrimination claim against solicitors Darling and Stephensons after being told, on announcing her pregnancy: 'I'm giving you notice today before you clock up job protection. You have chosen to become pregnant before the two years were up.'[30] It was because of the prevalence of such attitudes that the AWS established a Maternity Help Line in 1994.[31] The AWS was receiving many calls from women solicitors whose careers were being threatened by maternity, a reality borne out by EOC research which has detailed the differential and substantial impact of having children on women's employment.[32] Has the marriage bar been replaced by a baby bar?

It is not unknown for a woman solicitor, having told her employer of her pregnancy, to be advised to have an abortion.[33] Another was asked at interview whether she would consider having an abortion to further her career;[34] and in another instance the firm offered to pay for the abortion.[35] Other comments may not be so dramatic, but still exemplify the attitude of many employers, namely that pregnancy is a 'problem', rather than being seen as normal and unexceptional. Pregnancy is seen as problematic as it is measured against the male norm of employment patterns; the full-time worker, who works without any breaks, until retirement. Very few women's employment patterns are akin to this norm, and, because pregnancy necessarily involves a period of leave, it is often seen as an unjustified interference with the business of the firm, despite how short-term an approach this may be.

Often, women solicitors meet this attitude with great, and unfortunate, surprise. Understandably, many women perceive when starting their career that a firm which admires them and promotes their work will accommodate them when it comes to pregnancy, maternity leave and their childcare responsibilities. Unfortunately, what happens too often is that, on pregnancy, the firm no longer admires the woman and promotes her work, but sees her as lacking in 'commitment' to her work and her firm, and thus reduces the options open to her.

Commitment is always seen in terms of something which is lacking and is a term with a specifically gendered meaning. In order to be a successful and respected solicitor/partner, what is required is not just the requisite knowledge and skills, but also 'commitment' to the profession, firm and career. In her survey of senior and managing partners, Hilary Sommerlad found that commitment was named by virtually all employers as *the* principal criterion for selection for employment and for partnership.[36] However, the term 'commitment' is very vague and flexible and therefore is able to be moulded to suit any particular situation. The way in which commitment is interpreted often to women's disadvantage is exemplified by the statement of one managing partner that women had 'difficulty in long term planning caused by lack of ability to give commitment to their career'.[37]

Hence, childbearing and childcaring are seen as demonstrating a lack of commitment to a career, rather than a normal and unexceptional event which simply necessitates a period of leave, perhaps a change in working conditions for a further period, but actually does not necessarily effect any change as to the nature of the woman's responsibility to her work and career. Thus, although solicitors are often on leave for a variety of reasons, from illness to sabbatical, leave on account of pregnancy carries a stigma with it. Equally, becoming a father does not represent the deficit of commitment that is attributed to women on becoming mothers. Indeed, if anything, on becoming a father the man is seen as more responsible and more in need of employment and work opportunities. Indeed, in one Canadian study, it was found that having a family was an asset for a male solicitor, but a liability for a woman.[38]

The effect of this requirement of 'commitment' is that it can override the credentials of women solicitors which, as we saw in terms of their qualifications at law school and in vocational training, are generally better than men's. The requirement of commitment also means that it enables an interpretation of women's decisions to, for example, leave the profession as their 'choice'. In other words, women, by having children, choose not to have as much 'commitment' to their firm or career as men, and have therefore 'chosen' not to be taken as seriously by their employer. Thus, in terms of the human capital theory considered above, women's decisions to have children are painted as their choice, and in doing so they 'accept' that this reduces their 'commitment' and are therefore deserving of a lower status.

In addition, arrangements for leave and childcare are seen as the woman's responsibility, with little obligation on the part of a firm, father or society generally. Thus, when it comes to arranging maternity leave, and perhaps a change in working practices on return from maternity leave, women address these issues in terms of seeking to get a 'deal' done with their firm, supervising partner or employer. In other words, it is seen as the woman's individual 'problem' to sort out. This individualisation of childcare responsibilities is prevalent throughout society as little social responsibility is taken for childcare. This means that firms do not have to ask themselves, how, if they are demanding all these billable hours, are the staff going to arrange their family responsibilities? This is because it falls to the woman to arrange childcare as, hitherto, most employees have been men with partners who relieve them of any childcare responsibilities. The women employees have either not had any family, and have therefore been able to progress as they have adopted a standard (masculine) employment pattern, or, the women have sorted the matter out for themselves, almost seeking to pretend that they do not have a family, or the women have left. This is the impact of the dominance of the masculine career pattern.

Thus, until pregnancy is viewed as normal, from the firm's perspective, as illness or sabbatical leave or holidays, and childcare responsibilities are seen as a collective issue, the moment of conception will continue to exert a bar to women's career advancement. Only one per cent of women in the UK workforce are pregnant at any one time – this is not a large burden either on employers or the state.

Indicative of a firm's attitude towards pregnancy may be the extent to which its maternity policy exceeds the statutory minimum. A 1997 YWL survey found that 74 per cent of the largest 100 firms offered enhanced maternity pay to their assistant solicitors.[39] This is a significant improvement from the figure of 46 per cent two years earlier. However, the figure falls to 58 per cent for the largest 200 firms. The Law Society's 1997 surveys of firms of all sizes found that only one third of firms offered extended maternity leave to assistant solicitors, falling to only one quarter offering enhanced maternity pay.[40]

Equity partners of course are not employees and are not therefore entitled by law to any statutory minimum provisions. The YWL and Law Society surveys have found that where provision has been made for partners to take leave, the provisions tend to be more favourable than for assistant solicitors, but there remain a large number of firms which have no specific policy.[41] The introduction of a Law Society Practice Rule in 1995 should have gone some way to alleviating some of these concerns. The practice rule provides that if firms of solicitors have not adopted their own anti-discrimination policy, the model policy adopted by the Law Society will apply in lieu. The model policy states that the maternity rights of partners are to be no less favourable than those of employees.[42] However, this still leaves a number of problems. It is possible that a firm will have adopted an anti-discrimination policy which overrides this requirement without making any alternative provision. Even if the model policy is adopted, this will not solve all the issues regarding partners taking maternity leave which will need to be considered, such as future drawings, performance targets and the like. The woman partner will, therefore, still need to enter negotiations with her firm as to her entitlements, when she is not exactly in a strong bargaining position.

'Greedy' law firms

Law firms have been appropriately described as 'greedy' organisations in that they demand exceedingly long hours from their staff.[43] Thus, the publication of research showing widespread dissatisfaction with the long hours culture of the profession did not come as a surprise.[44] A later study found anxiety among assistant solicitors who were complaining about the long hours culture of law firms, and how they felt vulnerable in their jobs if they were not seen to be working the same hours as colleagues;[45] with another finding that more than half of those questioned found their hours excessive.[46] Most recently, a study of the attitudes of trainee solicitors found that women and men were searching for not just a fulfilling career, but a professional life compatible with a good 'quality of life'.[47] This long hours culture, having become entrenched, means that those who do not conform may be disadvantaged.

There are two particular aspects to this culture which may adversely affect women. First is the reality that women bear the brunt of domestic responsibilities and are therefore less likely than men to have the 'free' time or inclination to devote to 'presenteeism'. This can result in a double burden for women with children. Not only are they often seen to lack commitment by the very fact of their motherhood (see above), but this also means that they are less likely to meet the demands of the firm's long hours ethos.

The second aspect relates to the need, in some firms, for marketing and bringing in business, 'practice development', which means that working time extends into private time: private social life merges with business life. Thus, marketing demands extend the already excessive working hours. In addition, the culture of marketing throws up many different and difficult issues for women. Marketing very often revolves around certain typically male dominated interests, institutions and cultures which often exclude women explicitly or implicitly. The most explicit exclusion is the existence of all male networks through which contacts are made and from which professional work flows. It has been suggested, for example, that over half the solicitors in one northern town are freemasons, and this is the means by which work is gained and jobs handed out.[48] Less explicitly, but equally effective, is the dominance of masculine cultures in marketing initiatives, such as sport and drinking. One study of the habits of solicitors found that the drinking culture of marketing was an essential part of an apparently necessary bonding process between male clients and solicitors, a bond which appears to be essential to gaining and keeping work.[49]

This demand for social contact, a privileging of personal relationships over legal knowledge and skill, is becoming ever more important. An article in the *Law Society Gazette* in 1996 was headed: 'more law firms realise that the social skills of their fee-earners can radically impact on the bottom line'.[50] In particular, the article considered how the inability to socialise with clients, often involving 'moving seamlessly from the business meeting, to drinks and dinner ... could prove an obstacle to career advancement'. The fact that this is true does not lessen the fact that the socialising involved has a very significant gender element. Often it involves heavy drinking, in which women can and do engage. However, heavy drinking in a woman is (still) likely to be treated very differently from the heavy drinking of one of the 'boys'. For others, it can involve the attendance at all male clubs, with obvious difficulties for women. And, as women more often than not have primary childcare responsibilities, it may just be much more problematic for women to socialise after hours than for men.

In order to counter these exclusionary practices many women have sought to adopt strategies that better acquaint themselves with these masculine cultures. One woman partner of a Leeds firm told a conference in 1995 about how, in a bid to relate to her male clients, she switched her allegiance from rugby union to rugby league, and waded through a heap of car magazines to ensure that she was not at a loss on one of her clients' 'favourite subjects'.[51] Others have advised learning to play golf and how to bet.[52] Another woman partner has suggested that women solicitors must learn to 'recognise and then to exploit the differences between the sexes in a marketing context'.[53]

This emphasis on social skills and the need to conform to masculine cultures and values raises considerable difficulties. For example, what does it mean to exploit the differences in the sexes in the marketing context? Does this mean exploit one's sexuality in hope that sexual frisson will bring in the work? Isn't this simply conforming to a traditional expectation that women fulfill certain decorative roles, roles which emphasise their femininity at the expense of being taken seriously as a professional? But also, there is a limit to what can be achieved by the woman seeking to emulate her male peers. The 'marketing' which takes place after, for example, the golf, may take place in a clubhouse which excludes women. The woman seeking to understand the intricacies of rugby or football will not easily match the male colleague who has been imbued with sports culture all his life. The woman solicitor who invites a male client to lunch/drinks or whatever, as a marketing initiative, runs the risk of her offer being miscontrued. So long as it is a male club and culture to which women seek entry, they will only ever be honorary, not full, members.

In other words, women do not have the same 'cultural' capital that men have, no matter how much 'human capital' they possess. It is such 'cultural capital' which lies behind the privileging of certain individuals in the profession.[54] Thus, even if women invest in their careers to the extent of gaining the same credentials as men, and try to become more like men in their interests and social habits, they do not have the same opportunities.

Discrimination at the margins

What is significant about each of these factors discussed so far is their potentially cumulative effect. The woman who becomes pregnant is no longer seen as 'one of the boys', indeed she may no longer wish to go drinking with colleagues or clients, and this may begin a spiral of being undermined. A sexualised workplace, where (single) women are assets at the marketing activities, where sexual harassment is condoned or seen as a 'joke', where a macho culture pervades the corporate ethos, all combine to undermine and marginalise women, especially those who do not seek to conform to stereotype.

Indeed, there are a myriad of covert and unrecognised means by which an employer and its employees can effectively discriminate against women solicitors.[55] Such factors may include those mentioned above as well as control over the selection of cases handled, the degree of opportunity given for professional development, the distribution of secretarial, administrative and paralegal assistance, the input into management decisions, the degree of personal responsibility for cases given, the nature of questions asked at interview and the inferences drawn therefrom, the amount of access to clients and marketing opportunities, the extent to which mentoring is offered, and the distribution of employee benefits (monetary and non-monetary).

The impact of the cumulative marginalisation of women is that in all these subtle ways, an employer (and male superiors and colleagues) can greatly influence the workplace environment and career prospects of women solicitors. This influence can be to such an extent that, although the impact of any single act of discrimination may be insignificant and difficult to prove, the cumulative impact of such subtle acts may be large enough to 'legitimately' justify denying promotion and career advancement at a later stage. Such discrimination can be perpetrated with a low probability of detection, recognition and prosecution. Moreover, discrimination at these margins may have the effect of discouraging women from seeking promotion: it may be obvious to them that the odds are stacked against them. Women's options, choices and work environment are therefore constrained. It is not their 'choice' to be sexually harassed, to be marginalised in the workplace, to be excluded from many essential elements of building professional practice, and it is not their choice to be treated as signaling the end of their careers simply by becoming pregnant.

At root is a culture which is masculine, and in which women are not as highly valued as men. They are seen to lack authority and are always seen as 'other' than the male norm. Allied to this is the fact that to succeed one has to 'fit in', which often equates to being 'one of the boys'. This may have the most serious repercussions in terms of partnership decisions. Achieving partnership will often be the pinnacle of a legal career, but reaching that position for women is fraught with many obstacles. A partnership, as the very term suggests, is supposed to be based on mutual trust and confidence. As partners have unlimited liability for the debts of the partnership, they need to be able to rely on their partners. Thus, like all such groupings, whether it be a board of directors, the university committee or the law firm partnership, people appoint people like themselves, and the organisation becomes self-perpetuating. Thus, the partners appoint (clone) in their own image, and of course it is much more likely to be a male than female image.[56]

TOWARDS EQUALITY

How can one begin to redress these constraints, attitudes and discriminatory practices? A profession which is outwardly committed to justice, fairness and equality, cannot deny a large proportion of its membership the benefit of those principles. In order that such principles are fulfilled in more than formal terms, the culture of the workplace must change, and recognition given to the general nature of women's disadvantages and different needs, particularly in respect of pregnancy.

Equal opportunities policies

In 1995 the Law Society adopted a practice rule which provides that solicitors must not discriminate on the grounds of race, sex or sexual orientation in their relations with clients, staff, solicitors, barristers or other persons. All firms are required to adopt anti-discrimination policies, and where they fail to do so, the Law Society's model policy applies in lieu.[57] In addition, targets are set for the employment of ethnic minorities in solicitors' firms which specify that small firms (between six and ten fee earners) should have at least one ethnic minority fee earner, and smaller firms should take this policy into consideration. Large firms (11 or more fee earners) should have at least ten per cent of their trainees and five per cent of their fee earners from ethnic minorities.[58]

The Law Society's survey of law firms in 1997 found that only two fifths of firms had a written equal opportunities policy in place, of whom two fifths used the Law Society's policy, a further one fifth had modified that policy in part, and 35 per cent had written their own policy.[59] These results raise a number of concerns. First, 60 per cent of firms did not have a written equal opportunities policy. Although the Law Society policy will apply in lieu, this is not adequate. Few staff are likely to know that the Law Society's model policy applies to them. The lack of a policy is suggestive of a lack of commitment from the firm towards equal opportunities. No opportunity has been taken to involve the members of the firm in any consultation, educative or training processes, all of which form a vital part of the role of an equal opportunities policy. Even for those firms which have adopted a policy, the question arises as to the nature of it. A policy which is nothing more than a few lines stating that the firm is committed to equal opportunities is hardly worth the paper it is written on, especially if it is tucked away at the back of a staff handbook. Equally, the firm which has adopted a policy, but which has taken the opportunity to water down the provisions in the Law Society's model policy, is not to be commended.

The adoption of a policy, however detailed and progressive, is only the start.[60] To be effective, it needs to begin to exert cultural change in an organisation and this is only possible if a commitment to the policy is made at the highest level. In addition, it must be monitored and reviewed, staff need to know of its existence, that it is taken seriously by management, and there needs to be training of staff involved in the implementation and monitoring process. Involvement of staff at all levels in the drafting of the policy, giving employees a feeling of ownership over its content, will also aid the development of the policy and its implementation. When *The Lawyer* newspaper was seeking information about the equal opportunities policies of firms, it was reported that many were very coy about disclosing any information, and very few had any form of training for partners to recognise and deal with racial or sexual harassment.[61]

A crucial part of an equal opportunities policy, or comprising the subject of a separate policy, relates to the pay policies of the firm. Pay policies (if indeed they exist) in most firms are shrouded in mystery and it is all but impossible to find out what pay rates are across a firm. Add to this the increasing use of performance related awards and it seems that transparent pay scales are like winning lottery tickets: few and far between. In the light of the damning evidence of pay discrimination within the profession, the President of the Law Society and chair of the EOC wrote in 1997 to every solicitors' firm suggesting that they investigate their pay practices.[62] In particular, the letter suggested that firms consult the EOC's Code of Practice on Equal Pay. The Code urges employers to adopt an equal pay policy, carry out a pay review and take action to deal with pay inequality and has legal effect to the extent that it will be admissible in evidence in any proceedings brought under the Sex Discrimination or Equal Pay Acts.

Every firm of solicitors which has not already done so, must adopt, implement and monitor an equal opportunities policy, including an equal pay policy. In the words of an editorial in *The Lawyer*, it is time for law firms to 'practise what they preach'.[63] The solicitors' profession continues to be quite happy to advise others on appropriate employment practices, but seems to have an almost total inability to apply those principles of good practice to itself. It must be hammered home to lawyers at the earliest stages of their careers that if the guardians and advisors of the law are prone to abusing it themselves, it reflects badly on the profession as a whole and therefore the confidence within which it is held by the public.[64]

Flexible working practices

It was noted above that the solicitors' profession, like so many workplaces, is wedded to a particular employment pattern, derogations from which are rarely entertained. However, it would be wrong to suggest that there is no change afoot. The long-hours culture has come in for particular criticism, with one newspaper headline in 1996 stating: 'Our learned friends are stressed and depressed'.[65] The dominance of this culture of 'presenteeism' has led to many campaigns in recent years, both within the legal profession and beyond, which seek to improve working conditions for both women and men. The Young Solicitors Group has campaigned in recent years for changes in the profession under the banner 'quality of life', and the organiser of the 1998 Woman Lawyer conference, barrister Margaret McCabe, insisted that seeking change for women in the profession was entwined with achieving 'quality of life' for the individual.[66] Thus, the quality of life campaign is intended to improve the working lives of women and men, but is also linked very closely to the kinds of demands women have been making for many years.

Steps towards a greater 'quality of life' may involve what might generically be termed flexible working practices. A working environment has to be moulded where not only are women and men able to avail themselves of such policies, but that they can be followed without committing professional suicide. For flexible arrangements to work, the firm's prevailing culture must endorse flexible work arrangements as part of the mainstream career patterns for both women and men. This demands changes in attitudes, particularly male attitudes.[67]

Flexible partnerships

A step down this road was taken in 1997 by City firm Linklaters.[68] A change was made to its partnership agreement allowing partners (both women and men) to reduce their working time by as much as two fifths. The initiative began by looking at why there were so few women partners (ten per cent) compared with the intake of trainees (50 per cent women). The firm's research identified that its women assistant solicitors certainly performed as well as its men, and that although some women were leaving to look after children full-time, most were moving to other jobs in part because they perceived there to be more flexibility. It was also recognised that although there was already a policy on flexible working at assistant level, very few took this up as it was perceived that this was a dead end. Diana Good, the partner who engineered the adoption of the policy, discovered that women seemed to be deciding to leave the firm before they came up for consideration as partners as there is a perception of a glass ceiling. The firm wanted to do something to retain top lawyers (both women and men) in the long term.

The scheme established by Linklaters means that the individual partner will have to establish a business case for the flexible arrangement; it will have to suit the department, and the client needs, as well as the individual and it will be reviewed annually. What is fundamental is that there is a policy in place, and it is not just left to the ad hoc determination of individuals. This was seen as vital in that if the policy was to work in encouraging women to remain with the firm, there needed to be a specific policy in place so that women, and men, could see that there was scope for flexibility. Of the thirty women partners at Linklaters, four are on a flexible programme.

Diana Good spent over a year consulting both internally (with partners and assistants) and externally with other firms, institutions and clients. 'What I discovered', says Diana, 'was that the more you talk to people and think about the subject, the more willing everyone is to consider flexibility. The immediate reaction tends to be that it is impossible to combine any form of flexibility with providing a service to clients. But there are a number of times when clients cannot get hold of their lawyer because they are working for another client, away on business, or even (sometimes!) on holiday.' Moreover, there is no reason to suppose that the lawyer with a flexible arrangement is suddenly going to abandon the client at 5.30pm. With organisation, discipline and flexibility on both sides, Diana says, it should be possible for women and men to achieve a better balance and still produce a first class service to clients. However, it seems that it was the negative, knee-jerk response that characterised the reaction of many other City firms to Linklaters' actions.[69] Nonetheless, as the dust has settled, it seems that Linklaters has started something of a trend as other firms re-evaluate their policies.[70] Other firms should follow suit: Linklaters' example shows that flexibility can work, and can co-exist with fee earning and career success.

Nonetheless, such messages have yet to be heard in most firms. The Law Society's survey of firms of all sizes found that only 16 per cent of firms would allow flexible working for partners, 20 per cent stated that it would definitely not be allowed, with the rest remaining sceptical.[71] These reactions reflect the closed attitude of many firms who seem to consider that nothing needs to change. One firm said that it did provide some flexible working for assistants, but the difference with many such options is that they are in reality, what are called in the US on the 'mommy track'. This means that they are not on the partnership track and do not involve client work, but usually are involved in client and firm know-how. While such work is suited to many, it cannot be expected to deliver opportunities for all women.

Part-time work

There does seem to be some general movement towards allowing part-time work, although it is not clear whether firms intend this to mean fee earning work or not. The YWL 1995 survey of the largest 100 solicitors' firms found that two out of five said that they would allow partners to work part-time, although only one third actually had a part-time partner.[72] Four out of five stated that they would allow assistant solicitors to work part-time, and again fewer actually employed part-time assistants.[73] Although these figures demonstrate an element of acceptance to some flexibility of work, they also need to be studied carefully. If four out of five firms allow assistants to work part-time, but only two out of five allow partners to do so then; working part-time is effectively a bar to promotion. Thus, the chances are that the work that the part-time assistant is doing is not the high status work which might lead to partnership, and may be servicing or 'know-how' work referred to above.

It seems that this message is not lost on most women, as the 1995 survey found that only seven per cent of women partners worked part-time, and only four per cent of women assistant solicitors did so. In the 1997 survey, the number of firms responding that they would allow part-time working actually fell.[74] Furthermore, in 1995 only three-quarters of firms stated that part-time working would be available to men,[75] and, although the 1997 survey did not differentiate between part-time working for women and for men, one respondent added that part-time work might be available for women if it fitted with 'business needs', but 'not at all for males'.[76] This reinforces the notion that women are the primary carers of children, and that, although a special case may be made for them to work part-time, men would not be allowed to do the same thing. This reinforces the gender roles which hinder women in the work force. Equally, they inhibit men from taking a greater role in childcare, a development which is essential if women are to be enabled to achieve their full potential. The cultural concern for men, which has to be overcome, is that part-time work is intimately connected with the fact that in society at large it is women who work part-time, and generally in low status and low paid jobs. Thus men taking part-time work are crossing the gender divide and threatening their own masculinity.

Paternity leave

A number of policies can facilitate and recognise a greater role for men in childcare and childrearing. The provision of paternity leave recognises that a father is entitled to some time off work at the birth of his child; which is vital to allowing a more fulfilling family life for men, and greater possibilities for women. Thus, the provision of paternity leave, and not just on an ad hoc basis, does send signals regarding the attitudes of the firm. In 1995, just under one quarter of the largest 100 firms offered paternity leave, ranging from two to five day's paid leave.[77] By 1997, this figure had at least risen, but only to 39 per cent of the largest 100 firms, and 30 per cent for the largest 200.[78] The Law Society's 1997 survey of all sizes of firm found that one third of firms stated that they would allow paternity leave for partners, but only a quarter would grant assistant solicitors the same opportunity.[79]

Parental leave

A parental leave policy generally provides that women and men may take leave for a set period at the time of the birth or adoption of their child, and for women this is in addition to their maternity leave. Such leave is common in many continental countries with Sweden offering twelve months leave on 75 per cent of normal salary.[80] Parental leave is an important means by which women and men can be involved in family life. Encouraging men to take a greater role in the care and nurture of children, by allowing, for example, extended leave on the birth of a child, is crucial to enabling women to remain in the workforce without the added burden of the primary responsibility for childcare.

Unfortunately, parental leave policies are extremely rare in this country.[81] However, this situation will change to a limited extent in the next few years as the government must implement an European Community directive providing for three month's unpaid leave for women and men on the birth or adoption of a child. The fact that the leave is unpaid will of course seriously limit the potential impact of the directive, but it is at least a start in the right direction. With time and pressure, firms and organisations should, as with maternity leave, begin to offer parental leave over and above the statutory minimum.

The 1997 YWL survey sought information from the largest 200 law firms as to whether they had policies providing for parental leave.[82] Despite the fact that this directive has been adopted and will need to be implemented in the UK within the next few years, and that these were law firms which were being surveyed, the ignorance expressed as to the concept of a policy such as parental leave was worrying. Only eight per cent of the largest two hundred law firms offered any form of parental leave.[83]

This lack of flexibility on the part of law firms is not of course unique, and is replicated throughout workplaces. In defence of some firms, one study in Manchester and Liverpool suggested that the main reason why those firms did not offer flexible working arrangements was that they had never been asked to provide them.[84] However, what this signals is that those firms have not considered the question of flexibility and the conditions of their staff. This means that until there is a woman willing to put her head above the parapet, the firms will do nothing. In 1979, the Royal Commission on Legal Services recommended that the profession should consider arrangements enabling part-time work and work from home.[85] Twenty years later this recommendation must be restated, as it is imperative that firms consider a variety of working arrangements, especially in the light of the technological advances of recent years.[86]

In-house and sole practice: is the grass greener?

It was noted above, and in the previous chapter, that more women than men practice in-house, with the perception that this offers a more hospitable environment for women. Solicitor Denise Kingsmill who, prior to becoming deputy chair of the Monopolies and Mergers Commission, was a partner with City firm Denton Hall, has suggested that the way City firms treat women lawyers is a 'disgrace'.[87] She said that many law firms were badly managed with a 'macho atmosphere based on a long hours culture and a competitive rather than a co-operative environment', which is not a place where many women flourish.[88] It is also often suggested that the in-house solicitor, whether in the public or private sector, is generally afforded better working conditions. Supporting this perception is a 1991 study which found that women's salaries in industry were drawing level with those of men.[89] Thus, perhaps the pay inequalities of the private practice firm are not as evident in-house.

Statistics from 1997 show that the number of in-house legal departments is growing as companies are handling more work in-house rather than tendering it out.[90] Not only are the number of in-house lawyers increasing, but the proportion of women is also with one survey finding that 21 per cent of heads of legal departments were women.[91] Ironically, however, if this growth is fuelled by women leaving private practice, the private practice law firm is likely to become ever more male-dominated. Alternatively, on a more positive note, if there are more women seeking legal services from firms, perhaps this will exert change.

It has also been suggested that just as more women move in-house, more are also becoming sole practitioners. In 1997, one fifth of all sole practitioners were women, a proportion which has been steadily increasing.[92] Evidence suggests that while men move into sole practice as a result of redundancy or failing partnerships, women make a much more positive career choice, seeking control over their working environment.[93]

Evidence therefore implies that women are seeking a less restrictive and male-dominated working environment. However, working in-house or in sole practice is only likely to ever be possible for a small number of women. The lesson is that private practice firms must change their practices in order that women are able to achieve equal professional status in whatever their chosen area of work and practice.

CONCLUSIONS

The solicitors' profession is an extremely diverse community with often very little in common between the large City practice and the two partner high street firm. It is, therefore, very difficult to generalise about the experiences of those who work in the profession. However, what does appear to be a common thread throughout the profession is the undervaluing and marginalisation of women solicitors; what differs is the way in which this is achieved. Ultimately, it is attitudes which need to change: attitudes to pregnancy, and to the possibility of pregnancy, to the family responsibilities of all staff, to sexual harassment, to women in authority and to the conduct of relations with clients. Steps towards changing attitudes begin with changing work practices and patterns, including the introduction of greater flexibility and the recognition of men's family commitments. Ultimately, therefore, change will be of benefit to both women and men. Indeed, in view of the fact that men remain the decision makers in most firms, change will need to come from men. There is some evidence, from the 'quality of life' campaigns and the like, that men are beginning to re-evaluate their priorities and seek greater flexibility, but are reluctant to voice such desires. We must create an environment in which women and men can reconcile their work and personal lives without committing professional suicide.

PERSONAL TESTIMONIES OF WOMEN IN THE SOLICITORS' PROFESSION

Natalia Siabkin, freelance solicitor

Natalia has practised as a solicitor in a number of legal aid/high street firms, working in a broad range of fields including family law and personal injury work. She describes below her experiences of legal practice and how she has now found the work that suits her best, as a freelancer ...

'I am a first generation lawyer in my family. My parents are both immigrants, one Italian and one Ukrainian, who settled in London, worked hard and reared one daughter in relative working class comfort. After three A levels (English, History and Latin), my personal desire would have been to do either an English or Classics degree. However, I was aware that a degree did not necessarily guarantee work and I took the pragmatic decision to study law at university as a conduit to a career.

After six weeks into the law course at University College London, where Roman Law was compulsory in the first year, I was seriously doubting my enthusiasm for legal study. At that time at least fifty percent of the year were female and I certainly did not want to be viewed as anything less than a serious student, particularly after one other female fresher told me on the first day that she was only studying law to find a husband. She disappeared after the first session and I do not know whether she succeeded on either the legal or matrimonial fronts. I did manage to keep myself sufficiently motivated to complete the degree, including a thesis on Consumer Protection in Roman Law, from which I learned the derivation of the axiom 'caveat emptor'. This has occasionally proved useful.

I found articles with relative ease back in 1982 in a small three partner firm in Paddington. I cannot recall how many applications I sent out, but I remember going to a number of interviews and trying to give the answers I thought were expected. It was just as well that I did not hanker after a commercial City career and, I remember repeatedly stating that I wanted to do law for the individual. I had done my share of holiday work in various firms from the Holborn offices of a well known City firm (where I saw that articled clerks had very little responsibility) to a general high street practice in Catford. This confirmed my prejudices in favour of what I saw as real life legal practice, basically dealing with the problematic relationships of individuals, whether in the context of families, employment, housing etc.

From day one of my training I was taking instructions from clients and running files in front line situations. The firm had a canny system of utilising the articled clerks as a self-perpetuating legal aid department, in that there were generally four of us (mainly female) six months apart in experience and we would learn from and train each other covering a wide spectrum of litigation from injunctions and divorces, to consumer claims, landlord and tenant actions and crime, and we would also be available to assist on the partners' cases and do clerking duties.

Upon completion of a training which admittedly equipped me well to stay afloat in most 'sink or swim' situations, I realised that my status could only truly undergo metamorphosis into solicitor adulthood, if I moved on to another firm. Again, I found a position fairly easily in a medium sized firm in Uxbridge working as a solicitor in their litigation department. In this firm, I faced two personal realisations; the first was that I did not want to specialise too early in my career; and secondly, I became acutely aware of the male domination of the firm and the seemingly consequent territorial and competitive locking of horns euphemistically labelled 'office politics'. The other facet of this was the impenetrable and ritualistic phenomenon of male bonding both between solicitors and between solicitors and clients.

Although I was well regarded, I was resisting pressure to specialise and, after some two years, I found that I needed to return to my roots in a high street practice where I could again take up a more general lawyer's role for the individual. I then worked with a sole practitioner local to my home for some four years in which I had a fairly autonomous role running the litigation 'department'. This relative autonomy and the joy of being able to walk to work did for a while distract me from the issue that I actually needed a more demanding working environment.

I then worked with another sole practitioner, female this time, in a dedicated legal aid family law practice. This was a challenging position in that I took on a role with many hats from lawyer to staff and office manager. More than any other job this one tested my endurance and, I realised that much as I found an almost masochistic fulfilment in the psychology and practice of family law, it has to be positively unhealthy for the psyche to do more than 80 per cent of this type of front line emotionally battering work full time. I was then working long hours and taking work home including over weekends; I was forever running like a hamster in a treadmill to stay at least a respectable margin behind the demands of the work; clients who lived with their personal problems wanted me as their lawyer to share that personal involvement which is emotionally impossible if not dangerous - both in terms of overload and in terms of losing objectivity. I worked in a highly charged and stressful environment, which also caused knock on friction in my personal life.

When I finally took a step back from my situation, and was unable to trace any ongoing job satisfaction, I decided that I had to get my enthusiasm levels back into credit. I had accepted that in my chosen area of legal practice monetary remuneration was never going to be awesome but I needed some compensation in terms of job satisfaction and quality of life outside work. For a while I made a sideways move taking up a position in a firm outside London, where one of the two partners was a personal friend (female) and where there was an opportunity to break into a more commercial field, though I found that issuing writs for finance companies did not regenerate my enthusiasm for the law. The personal friend is a very successful lawyer and, with a great deal of effort has managed to combine a working and family life, though there were ongoing stresses between the two competing interests.

In any event, I had reached mid-career crisis and, needed a period of reflection. I knew that amongst my contemporaries some had left the law and others were very disillusioned with legal practice. It did cross my mind that I would also leave the law and, perhaps pursue a literary career (as I am sure many others before and after me have fantasised) but in the meantime I needed to pay my bills and fell into freelancing. Surprisingly, I had accidentally found a type of legal practice which suited my then needs personally and professionally. I discovered the relative freedom of being self-employed with minimal and controllable overheads. The future is unpredictable but this has its own potential opportunities and appeal.

After signing up with the market leader in locum recruitment, I deliberately set my sights on very short term assignments; within a day or so I was working on a week to week basis in a Fleet Street practice doing mainly family and personal injury litigation. This lasted for about six weeks and was followed by a stint in a small niche practice in Bow Lane. I was then offered an ongoing two day per week position doing civil litigation (mainly personal injury) in a firm specialising in criminal and mental health work in Charing Cross. After this I answered an advert in the Law Society Gazette for a maternity leave assignment in a small reputable family law practice. Since then I have been able to get further work on recommendation, including another six month maternity leave cover for a high profile matrimonial lawyer in a prestigious firm. In between work, I have travelled, including a memorable and inspiring trip to Peru where I walked the Inca trail to the lost city of Machu Picchu. I also spent six glorious weeks in Tuscany in Italy, and for some of that time I studied Italian in Siena, reclaiming my heritage.

As a freelancer I find that I have to combine the skills of cautious diplomat, amateur psychologist and firefighter, as well as keeping my legal knowledge on the ball. The attractions of freelancing to the individual are flexibility, variety and skimming over the leaden side of office politics. The attractions to the employer are finite costs and hopefully, an efficient professional service on tap.'

Caroline Graham, City solicitor

Caroline has practised in three different corporate-commercial firms and is presently based in the City. She describes below her experiences of this type of work and her views on discrimination in the profession ...

'In the mid-1980s, a certain comprehensive school in the north-east of England had what could almost have been seen as a policy; bright kids were encouraged to apply to read medicine. If science A levels did not seem to represent their best chance of gaining an Oxbridge place, then it would have to be law instead. So in 1987, armed with A levels in English Literature, French and Maths, I went to Cambridge to read law.

Of the ten students in my year at Girton College, nine of us were women and the Director of Studies was the inspirational Dr Stephanie Palmer, one of a few beacons in my university years. Stephanie was one of the few Cambridge academics who spoke my language - feminist, strong-willed, and interested in law in its social context rather than as the rules-based discipline so often taught at Cambridge. But these were the yuppie years, and Cambridge was something of a 'sausage factory'. Most of the law graduates in my year went straight into articles at City firms. I knew nothing about business, and I had entirely run out of money; becoming a management consultant seemed a way to kill two birds with one stone.

Two years later, having saved enough to live comfortably for a year, I went to the College of Law, and then to the City firm Herbert Smith, where I found I enjoyed being a 'sausage' more than I had expected. It is easy, as a student, to assume that 'law in a social context' must mean criminal law, family law, perhaps human rights, and had I made my career choices straight from University I might have made the same mistake. But our society, in common with most others in the world, is underpinned by commerce and trade, and as a corporate lawyer you are part of the system which regulates and controls the capitalist economy, which is so fundamental to our way of life.

I did not remain in the City when I qualified. My personal circumstances enabled me to move back to the north east of England, where my family and my roots remained. I joined the corporate department of a large Newcastle firm, to combine work on corporate transactions with corporate tax. Two years later, I returned to London. Partly, this was because I love London, with its racial and cultural diversity; partly it was because almost all top quality work (however loudly provincial firms protest) still goes to London. But in large measure, it was because I found my old firm to be an inhospitable place for a woman.

The 'Sexist North' may be a stereotype, but I found a degree of discrimination which I had not encountered before. One partner (in his thirties, up-ending the common view that dinosaur attitudes will soon become extinct) told me over lunch that he had 'never met a good woman lawyer'. Eight new partners were made up in the last year, but all were men. Ogling and inappropriate remarks on occasion astonished me and there was anecdotal evidence of pay discrimination.

Firms need to recognise that they will lose high quality, expensively trained staff if they continue to ignore direct and indirect discriminatory practices. I was the fourth woman to leave my department in under two years and a belief that 'a woman has to be better than a man to succeed here' was mentioned to me as a reason by one of these women. It is just as important, though, that firms recognise that recruitment, as well as retention of good female lawyers, will become more difficult unless these issues are addressed.

It would be wrong to suggest that discrimination - on grounds of race, age, sexuality and disability as well as sex - has been eliminated in the law. It is sometimes said that lawyers, like the police, enforce the law but do not always show the same commitment to its application. Even as a trainee solicitor, one is aware that women are asked less often to take part in marketing initiatives, which are often focused on sport. Since it is in part through marketing that one develops client relationships, and thereafter builds a practice, this is a more important area than it appears.

I do not have a family and so I do not have first-hand experience of the discrimination which, I am told, accelerates at this time. However, it is certainly true that success in part follows perceived commitment, and women are often regarded as less committed than their male colleagues when they become parents.'

Mary Dowson, Principal Legal Executive, Middlesbrough Council

Mary is a legal executive with Middlesbrough Council undertaking a wide range of advocacy and litigation work. She explains below her route to the law and the nature of her work ...

'I was born in Middlesbrough in 1962 and came from a very hard working, determined and disciplined family background. I attended secondary school in Middlesbrough and after obtaining my O levels decided to pursue a career in business and commerce, and I was also determined to take further higher qualifications at college.

I obtained a position with a local authority, Middlesbrough Borough Council, in 1978 and started my career as a Trainee Administrative Clerk. I attended Kirby College and obtained further higher qualifications, equivalent to A level standard, in public administration, over a two year period. As part of these qualifications, I studied law and knew at this point in time I very much wanted to become a lawyer.

Fortunately, working for a local authority in the public sector enabled me to obtain a position as a Legal Clerk in 1981. I had researched the different routes on how to become a lawyer and chose a route through the Institute of Legal Executives. The main reason for my decision was the possibility to work within a legal office and obtain law qualifications on a part-time basis. To me this flexible approach seemed to be the best option: learn and earn. This route would ensure that, by the time I had acquired my qualifications, I would also be a very experienced lawyer.

It took over four years to obtain my qualifications, and I then had a further two years' practical experience in order to acquire the status of Fellow of the Institute of Legal Executives. Since qualifying as a legal executive, I organised the creation of the North East Branch of Legal Executives. I was elected chairperson of that organisation in 1993 and currently hold the position of vice-chairperson. In 1995 I was elected as a National Council Member and director of the Institute of Legal Executives.

In 1986 I obtained a position with Middlesbrough Council as a Legal Executive specialising in the field of litigation. I was later promoted to the position of Senior Legal Executive, specialising in civil litigation and criminal matters, for Langbaurgh-on-Tees Council. A year and a half later I obtained further promotion as a Principal Litigation Officer with my old authority, Middlesbrough Council. At this point in my career, I managed a civil litigation section with a team of legal executives and legal clerks and carried out various duties as one of the Council's Advocates, including assessing evidence in respect of criminal and civil cases, appearing on behalf of the Council in both civil and criminal cases in respect of not only uncontested matters but complex contested matters including a variety of trials, appeals and arbitrations.

It was possible for me to acquire this advocacy experience since working within the public sector gave me greater rights of audience as a legal executive. It is essential at this level that you have the ability to communicate effectively cope with pressure, recognise and solve problems as they arise, and present cases with confidence and clarity.

In 1996 I obtained further promotion and presently occupy the position of Principal Legal Executive with Middlesbrough Borough Council. I enjoy my work immensely and my current duties include mostly advocacy and managing a Litigation Section. I have instigated many changes within the Division to adopt a much more business-effective approach to legal services and I am heavily involved with our own Legal Department's Compulsory Competitive Tendering (CCT) bid for legal services and the implementation of a case management system to ensure best value is achieved. I also advise Client Departments on a daily basis on CCT regulations and European Public Procurement Legislation. In addition, I am also responsible for the training and evaluating all legal executives and other legal staff within the Division, and I am responsible for the updating and overseeing of information technology within the department.

Throughout any legal career within the public or the private sector, you are always going to be faced with obstacles and will have to overcome experiences such as discrimination. You must be determined and strong enough to achieve your goals and objectives no matter what problems may arise. In local government there are active policies in place against discrimination on the grounds of sex, race, creed, origin or disability and also policies for equal opportunities and a fair selection and recruitment procedure. My employers have also recently signed a letter of intent with regard to the investment in people and have made a public statement committing themselves to the development and training of its workforce. I am not saying that discrimination or equal opportunities matters do not arise in the local government domain and I certainly have had my own fair share of these experiences as I have progressed through my career. However, as long as you know your rights and you have the ability to communicate effectively with your employer and other organisations and encourage procedures and practices to be operated in the working environment, you can overcome these hurdles.

I have always believed that women do have to try harder to achieve their ambitions in law. However, as long as you are determined enough you will achieve it. With regard to reform measures and future prospects for women in the law I do believe that to achieve your ambitions you do have to work hard. However, you are also entitled to a 'quality of life'. Employers and professional institutions should appreciate this in all types of professional careers and consideration should be given to changing working practices ie introduction of flexible working hours, part-time appointments, temporary cover for maternity leave. It is one of my firm beliefs that if you recognise the quality of the staff you have and you want to keep that quality you should ensure that needs are catered for within the workforce, and this applies equally to women and men.

I would advise any woman contemplating a career in law to take a look at the route that I chose and consider qualifying as a Fellow of the Institute of Legal Executives (ILEX). ILEX has developed an education and training scheme which offers the opportunity of a career in law to everybody. Legal executives are qualified, experienced lawyers specialising in a particular area of law. They have their own recognised status and role within the legal profession and enjoy a responsible and rewarding career. The training scheme leads to a professional qualification in law and also legal practice. ILEX's equal opportunity approach enables mature students to obtain the qualification and also offers long distance training. I would certainly not change the route I had taken to qualify in the law which I now find so challenging and rewarding.'

NOTES

1   Lord Finlay was speaking against a bill to allow women to enter the solicitors' profession: Hansard (Lord's) XXIV, 267–8, quoted in Hilary Sommerlad and Peter Sanderson, *Gender, Choice and Commitment*, Dartmouth-Ashgate: Aldershot, forthcoming.

2   Concerning Solicitors, Chatto & Windus: London, 1920, p 46, quoted in Mary Gaudron, '*Speech to Launch Australian Women Lawyers*', 72 (1998) *Australian Law Journal* 119–124, p 120.

3   See Deborah Rhode, 'Perspectives on Professional Women', 40 (1988) *Stanford Law Review* 1163–1207, pp 1166–1167.
4   Ibid p 1167.
5   Bill Cole, *Trends in the Solicitors' Profession – Annual Statistical Report 1997*, Law Society: London, 1997, p 2. All statistics referred to in this chapter are taken from this report unless otherwise stated.
6   26 per cent of women on the roll do not hold a PC, compared with only 20 per cent of men: see Verity Lewis, *Trends in the Solicitors' Profession – Annual Statistical Report 1996*, Law Society: London, 1996, p 6.
7   Law Society, *Equal in the Law*, Law Society: London, 1988, p 9, ibid p 76 and supra note 5.
8   Clare McGlynn and Caroline Graham, *Soliciting Equality – Equality and Opportunity in the Solicitors' Profession*, YWL: London, 1995, p 8. See also: Clare McGlynn, 'Soliciting Equality – the way forward', 145 (1995) 1065–1066, 1070. For further information on YWL, see appendices five and six.
9   Supra note 5, p 17 and Law Society, *Trends in the Solicitors' Profession – Annual Statistical Report 1994*, Law Society: London, 1995, p 13.
10 Supra note 8, pp 8–9.
11 Clare McGlynn, 'Where men still rule', *The Times*, 22 April 1997.
12 Hilary Sommerlad, 'The myth of feminisation: women and cultural change in the legal profession', 1 (1994) *International Journal of the Legal Profession* 31–53, p 34.
13 Ethnic origin is known for 86 per cent of those admitted.
14 The Law Society's annual survey of solicitors' firms is seeking information on ethnic minorities, the results of which should be available by the end of 1998.
15 Discussed in Clare McGlynn, 'Paying for Equality', 147 (1997) *New Law Journal* 568–569.
16 See Law Society Press Release, *New Survey Reveals Continuing Inequality between Male and Female Solicitors*, 7 July 1997 and ibid.
17 For further information on the AWS, see appendices five and six.
18 *The Guardian*, 2 September 1997.
19 Quoted in Evlynne Gilvarry, 'Outspoken voice strives for top job', *Law Society Gazette*, 12 April 1995.
20 Quoted in 'Lucy Winskell ... Council Impressions', Young Solicitors Group Newsletter, Spring 1997.
21 Supra note 19.
22 *The Guardian*, 22 April 1996.
23 *The Lawyer*, 21 July 1998.
24 Martin Mears, 'Aux armes to defend the revolution', *The Times*, 16 April 1996.
25 For a discussion of this theory, and its application and assessment in the light of research on Canadian lawyers, see John Hagan and Fiona Kay, *Gender in Practice – A Study of Lawyers' Lives*, Oxford University Press: Oxford, 1995.
26 See Gary Becker, *A Treatise on Family*, Harvard University Press: Cambridge, Mass, 1991.
27 Michael Levin, 'Women, Work, Biology, Justice', in Caroline Quest (ed), *Equal Opportunities: A Feminist Fallacy*, Institute of Economic Affairs: London, 1992, p 14, emphasis in original.

28 See further, Margaret Thornton, *Dissonance and Distrust – Women and the Legal Profession*, Oxford University Press: Oxford, 1996, pp 10–40.
29 See further, Hilary Sommerland, 'The Gendering of the Professional Subject: commitment, choice and social closure in the legal profession', in Clare McGlynn (ed), *Legal Feminisms: theory and practice*, Dartmouth-Ashgate: Aldershot, 1998. See further below under the heading ' "Greedy" law firms'.
30 *Dowson v Darling and Stephensons*, case 2503908/97, IT, considered in 3 (1998) *Employment Lawyer* 3–4. It is the ignorance of the law, as well as the attitudes displayed, which make this case remarkable.
31 The help line is run by Judith Willis and can be contacted on 0181 676 9887.
32 EOC, *The Lifecycle of Inequality*, EOC: Manchester, 1995, pp 13, 19.
33 *The Guardian*, 8 April 1997.
34 Quoted in James Goldston, 'Pregnant Pause', *Legal Business*, 1992.
35 *The Guardian*, 8 April 1997.
36 Supra note 29.
37 Ibid.
38 Supra note 25, p 91.
39 Supra note 11.
40 Judith Sidaway and Gillian Davidson, *The Panel: A Study of Private Practice 1996–97*, Law Society: London, 1998 and Bill Cole, *Solicitors in Private Practice: Findings from the Law Society Omnibus Survey*, 1997, Law Society: London.
41 Supra notes 8, 11, 40.
42 The AWS has produced draft maternity clauses for partnership deeds which provide for six month's leave, three months on full drawings, three months on half.
43 Carrie Menkel-Meadow, 'Feminisation of the Legal Profession: the comparative sociology of women lawyers', in Richard Abel and Philip Lewis (eds), *Lawyers in Society: Comparative Theories*, University of California Press: Berkeley, 1989, p 239.
44 *Law Society Gazette*, 8 November 1995.
45 Roger Trapp, 'Quarter of female solicitors harassed', *The Independent*, 27 January 1997.
46 Dan Bindman, 'Assistants baulk at long hours', *Law Society Gazette*, 13 August 1997.
47 Chris Fogarty, 'Trainees who bluffed their way in, want out' and Ravinder Chahal, 'A choice of career or lifestyle?', both in *The Lawyer*, 19 May 1998.
48 Chris Mullin MP quoted in Jessica Smerin, 'Brothers in law', *Law Society Gazette*, 5 February 1997.
49 See further Hilary Sommerlad and Peter Sanderson, supra note 1.
50 Sue Stapley, 'Social Climbing', *Law Society Gazette*, 25 September 1996.
51 Quoted in Evlynne Gilvarry and Jessica Smerin, 'Market shy', *Law Society Gazette*, 12 April 1995.
52 Quoted in Jayne Willets, 'Getting in the business', *Law Society Gazette*, 26 March 1997.
53 Ibid.
54 See further Hilary Sommerlad and Peter Sanderson, supra note 1, and Hilary Sommerlad, supra note 12.

55   In the context of the US, see Bernard Lentz and David Laband, *Sex Discrimination in the Legal Profession*, Quorum Books: USA, 1995, discussed in Clare McGlynn, 'Sex Discrimination at the Margins', 146 (1996) *New Law Journal* 379–381. In relation to academics, see chapter three, under the heading 'Women on the margins'.

56   It is also just as likely to be a heterosexual, white and able-bodied image. If proposals to allow partnerships to become limited in their liability are adopted (see Jennifer Payne, 'Limiting the liability of professional partnerships: in search of this Holy Grail', 18 (1997) *Company Lawyer* 81–88), perhaps this will alter the culture of partnerships, becoming less of a brotherhood and more of a professional, arms-length, business relationship, possibly leading to the falling of barriers against women.

57   A leaflet explaining the practice rule and setting out the policy is available from the Law Society.

58   The Law Society's annual survey of solicitors' firms is presently seeking information from firms regarding compliance with these targets, the results of which should be available by the end of 1998.

59   Bill Cole, supra note 40.

60   See further chapter two under the heading 'The equal opportunities agenda'.

61   Robert Lindsay, 'You really wouldn't expect it of law firms', *The Lawyer*, 28 January 1997.

62   Law Society Press Release, *New Survey Reveals Continuing Inequality between Male and Female Solicitors*, 7 July 1997.

63   Editorial, 'Time to practise what is preached', *The Lawyer*, 28 January 1997.

64   Ibid.

65   *The Daily Express*, 6 May 1996.

66   Quoted in Frances Gibb, 'An end to the old sex war?', *The Times*, 3 February 1998.

67   See also Joanne Gubbay, 'My learned friends want to be more flexible', *The Times*, 14 April 1998.

68   See Robert Rice, 'Female attraction', *The Financial Times*, 3 June 1997.

69   Ibid.

70   For example, Clifford Chance, the largest solicitors' firm in the UK, has announced that it is considering introducing home working for its partners: *Law Society Gazette*, 7 May 1998.

71   Bill Cole, supra note 40.

72   Supra note 8, pp 10–11.

73   Ibid.

74   Supra note 11.

75   Supra note 8, pp 10–11.

76   Ibid.

77   Supra note 8.

78   Supra note 11.

79   Bill Cole, supra note 40.

80   See Equal Opportunities Review, 'Parental and family leave', 66 (1996) *Equal Opportunities Review* 22–29.

81   See further Helen Wilkinson and Ivan Briscoe, *Parental Leave: the price of family values?*, London: Demos, 1997.

82   Clare McGlynn, 'Time is ripe for parental leave', *The Times*, 27 May 1997.
83   Ibid.
84   Sue Nott, 'Women in the Law', 139 (1989) *New Law Journal* 749–750.
85   The Royal Commission on Legal Services, Cmnd 7648, 1979, para 35.24, p 500.
86   See further: Paul Hilder, *IT in the Solicitors' Office*, Blackstone Press: London, 1997.
87   Quoted in Neil Rose, 'In-house teams "attract women" ', *Law Society Gazette*, 4 February 1998.
88   Ibid.
89   Quoted in Eleni Skordaki, 'Glass slippers and glass ceilings: women in the legal profession', 3 (1996) *International Journal of the Legal Profession* 7–43, p 22.
90   Neil Rose, 'In-house teams in growth boom', *Law Society Gazette*, 9 July 1997.
91   Ibid.
92   Bill Cole, supra note 5, p 17. In 1994, 16 per cent of sole practitioners were women.
93   Rebecca Towers, 'Sole practice enables many women to combine work and family', *Law Society Gazette*, 16 April 1998.

Chapter 5

Practising as a barrister: the Bar Vocational Course and pupillage

'There is nothing quite like the feeling of personal fulfilment when you sit down after making a successful speech or concluding a devastating cross-examination.'

Geraldine Andrews, Essex Court Chambers

'There has never been a better time for a woman to come to the Bar.'

Laura Cox QC, Cloisters

'Being a barrister is a wonderful job. It is interesting, challenging and, if you're lucky, useful.'

Elizabeth Woodcraft, Tooks Court Chambers

Geraldine Andrews was attracted to study law and pursue a career at the Bar by the 'combination of the performance element and the academic challenge'. Dame Barbara Mills QC, Director of Public Prosecutions, was equally attracted to the advocacy of the Bar and 'also to the sense of fair play between parties who were in dispute'. Najma Khanzada became a barrister because she 'wanted a profession that reflected my social welfarist politics and, at a superficial level, I wanted to stride around in sharp suits and be paid for being aggressive'. For some of these reasons, and many, many more, over 400 women undertake the Bar Vocational Course (BVC) each year, most hoping to pursue a career at the Bar. This high number of women is a marked and welcome change from the years when women barristers were few and far between.

This chapter sketches the path which the prospective barrister will need to navigate to pursue her ambition to become a barrister, looking at the BVC, pupillage and gaining a tenancy. In this context, it goes on to examine questions of access to the Bar. The chapter concludes with the personal testimonies of three women barristers, which, together with those at the end of chapter six, expand on and exemplify many of the themes addressed in this and the following chapter.

GETTING STARTED: THE BAR VOCATIONAL COURSE

In order to qualify as a barrister, a student must undertake the BVC, having first completed the academic stage of training. Until 1997, the BVC was only offered at one institution, the Inns of Court School of Law in London. As a result of a number of pressures, including the perceived need to open the Bar course to those wishing to study outside London, a number of institutions now offer the BVC throughout the country. The aim of the BVC is to integrate the development of the practical skills required to be a barrister, together with the teaching of specialist areas of law, such as civil and criminal litigation and evidence. An application for the BVC must be made through the centralised applications procedure known as CACH (Centralised Applications and Clearing House).[1]

WHO WINS? APPLICATIONS AND ACCEPTANCES

The proportion of applicants to places on the BVC has increased over the years. In 1986, 80 per cent of applicants for the then equivalent of the BVC registered on the course.[2] The figure fell to 43 per cent by 1994–95, but it has increased to 67 per cent since then.[3] The institutions offering the BVC have stated that the principal criterion for admission is degree classification. This generally means that a student needs to have either a first class or 2.1 degree, or be predicted a 2.1, although in the end just under one third of students on the BVC have a 2.2 degree or less. As the number of institutions offering the BVC increased in 1997, there are now a further 1,310 full time places available. Whether this will result in a greater number of applications, or a lower applications-acceptances ratio, is not yet clear.

It is encouraging to see from table 5.1 that the number of women embarking on the BVC is increasing, although it has yet to match the equality of numbers to be found on the solicitors' Legal Practice Course (LPC). Unfortunately, there are no statistics available showing the number of applications made by women, and therefore it is not known whether the lower number of women on the course reflect fewer applications by women, or is influenced by some other factor. This is information which should be readily available, especially since the advent of the CACH centralised applications system.

In terms of access to the Bar, an important factor is wealth. The course fees for the BVC range from around £5,500 to £6,500, and, on top of the fees, students require accommodation and living costs, as well as financing books, study, and membership of and dining at the Inns of Court (see below). Not surprisingly, therefore, recent Law Society research found that of those who had been offered a place on the BVC, just under one quarter did not take up that place on the grounds that they could not finance the

Table 5.1: Number of women taking the BVC 1994–1997

	1996–97	1995–96	1994–95
Women accepted	436 (42%)	436 (38%)	366 (36%)
Men accepted	596 (58%)	702 (62%)	661 (64%)
Total accepted	1032 (100%)	1138 (100%)	1027 (100%)

course.[4] It remains the case that the greatest source of funding for those training to become barristers is parental contribution, but not everyone's parents are in a position to foot this bill. In addition, as noted in chapter three, there is evidence that women students earn less than men in so-called 'holiday jobs', which, together with the fact that women earn less at the Bar than men, may mean that the financial commitment falls heavier on women than on men.[5]

The Inns of Court offer scholarships and a variety of smaller awards for call fees, admission fees, pupillage and the CPE based on merit and ability. Also, the Bar Council administers the Bar Scholarship Trust which provides a minimum of nine interest-free loans annually. Local education authority grants are discretionary and their availability is decreasing, and in 1995–96 only six per cent of BVC students had their course fees paid by their local education authority, with a further seven per cent receiving a contribution towards fees.[7]

Thus, where the Bar Council's guide to becoming a barrister states that it is not only commitment to the Bar which is required, but 'financial commitment', this is very true, but equally very disappointing. Success at the Bar should not be dependent on whether an individual has access to substantial sums of money, whether given or borrowed. The progress which has been achieved in making the Bar more open is threatened by the increasing financial pressures, as is the Bar's reputation in society in the longer term. Barrister Josephine Hayes, who is Chair of the Association of Women Barristers (AWB), says that she 'could not have come to the Bar without a local authority grant'.[8] This is a sentiment echoed by the present Chair of the Bar Council, Heather Hallet QC, who says that the 'likes of me' would not have been able to pursue a career at the Bar without a local authority grant.[9]

Joining Inn

The costs of becoming a barrister do not stop with fees for the BVC. Before commencing the BVC, each student must join one of the four Inns of Court, which costs around £80. The Inns of Court are colleges of barristers dating back to medieval times, and only the Inns have the right to 'call' women and men to the Bar, that is, to admit them as barristers. There are four Inns: Middle

Temple, Inner Temple, Gray's Inn and Lincoln's Inn. The Inns have students' associations which organise mooting and debating competitions, as well as social events. The decision as to which Inn to join may owe much to the availability of scholarships in each Inn and/or any links which the Inn might have with a student's college or university.

The need to join an Inn stems from the fact that a student must eat 12 dinners, costing around £10 to 20, at their chosen Inn before qualifying as a barrister. At least six dinners must be eaten before the BVC examinations. The tradition of eating dinners was initiated at a time when all prospective barristers studied in London and the number of dinners to be eaten was only reduced to 12 in 1996, having been 24 a few years earlier. A recent Bar Council working party has agreed that there should be a higher educational content to dining, presumably as an attempt to justify the requirement to eat dinners.[10]

Some have argued that dining provides an environment in which barristers can meet and the novice can be 'instilled with the tradition' of the Bar.[11] No doubt many barristers and students enjoy eating dinners in the Inn, but to make it a professional requirement for qualification as a barrister is an absurd anachronism. Further, the fact that the dinners have to be eaten in London displays what has been called the Bar's 'unimaginatively metropolitan mindset',[12] and will add considerably not only to the cost of studying for the Bar but also to the sense of their being two Bars and two classes of barristers' training: one in London and the other not.

In her book, *Eve was Framed*, Baroness Helena Kennedy QC eloquently describes the obsessively traditionalist, archaic and often puerile nature of dinners in Gray's Inn when she was a student. Having been called to the Bar, Baroness Kennedy did not return to her Inn for ten years, expecting that in the meantime the influx of students would be much more egalitarian, only to be disappointed that the Inn was debating a motion something like: 'A woman's place is on her back.'[13] Although times have moved on, women remain almost totally absent from the ranks of the Benchers, who run the Inns, with only 41 out of 671 being women.[14] In 1996, there were 13 women Benchers in Gray's Inn, 12 in Inner Temple, eight in Middle Temple and eight in Lincoln's Inn.[15] Benchers are not elected by members of the Inn, but decide among themselves who they will invite to become Benchers. Many calls have been made for the reform of the Inns, which will be examined in the next chapter.[16]

Sex, ethnicity and passing the BVC

Examination results always excite controversy, and none more so than when those results demonstrate a difference on the basis of the ethnicity of the student. The overall pass rate on the BVC in 1995–96 was 77 per cent,[17] ranging from 58 per cent for ethnic minority students to 81 per cent for all other students. The failure rate for ethnic minority students in 1995–96 was in fact almost double that for 1994–95 (25 per cent).[18] The fact that ethnic minority students have a considerably lower pass rate to other students on the BVC has been a cause for concern throughout the 1990s. An investigation into the BVC in 1994, resulting in the Barrow Report, concluded that the most likely reason for the disparity in pass rates between white and ethnic minority students was the failure of the Inns of Court School of Law and the Bar Council to 'meet the full needs of black and ethnic minority students during the vocational stage of training'.[19] Analysis of the pass rates published by the Society of Black Lawyers (SBL) in 1997 showed that since the Barrow report in 1994, the prospects for black barristers had actually got worse. The SBL concluded that the figures represented an 'appalling level of unequal treatment', suggesting that black students were not getting the extra assistance that white students received, and ethnic minority students were also suffering in oral examinations which were subjectively judged.

Table 5.2: BVC 1995–96 pass rates by sex

	Women		Men		Total	
Outstanding	4	(1%)	10	(2%)	14	(1%)
Very competent	105	(24%)	180	(26%)	285	(25%)
Competent	202	(47%)	362	(52%)	564	(50%)
Fail	123	(28%)	140	(20%)	263	(24%)
Total	434	(100%)	692	(100%)	1126	(100%)
Overall passes	311	(72%)	552	(80%)	863	(77%)

The statistics for pass rates by sex are equally revealing and can be found in table 5.2. The only figures available are for 1995–96, and therefore no comparison with previous years can be made. Nor are any figures available which provide an analysis by both sex and ethnicity. Table 5.2 reveals that not only is the overall pass rate for women lower than for men, but the proportions by which women gain the highest awards are lower than those for men. This compares unfavourably with the fact that on average women are awarded more upper second and first class degrees at undergraduate level and have a higher pass rate on the LPC.[21]

MOVING ON: PUPILLAGE

'I can truly say that my pupillage was the most wonderful experience.'

Laura Cox QC

With the BVC completed, the next stage to practising as a barrister is what is known as 'pupillage'. During pupillage, the pupil – a trainee barrister –

will shadow the work of her pupilmistress or master and will gradually take on cases of her own. This period of training 'on the job' is often divided into two six month periods, known as 'sets' or 'sixes'. Pupils will often spend their first six in one set of chambers and their second in another, although it is also possible to spend twelve months in one set. In the first six, a pupil does not accept any briefs, ie instructions from solicitors, but in the second six pupils can begin to represent clients and can expect to start earning. It is increasingly common for pupils to do a third set of six months, if they are unable to obtain a tenancy straight away. The quality of training a pupil receives varies enormously between sets of chambers, but new measures are being instituted by the Bar Council in order to ensure minimum standards of training. The new system, involving visits to chambers, is intended to improve, and make more professional, the system of training in the hope that it may stamp out some of the discriminatory practices which arise during pupillages.[22]

Many pupils will already have some experience of working in chambers from mini-pupillages. A mini-pupillage can range from one day, to one week, to a few weeks, and such experience is all but a prerequisite to becoming a barrister. Depending on when the experience is gained, it can help make decisions about whether to study law at university, whether to go into the legal profession, whether to become a barrister and the area of practice in which to specialise. For the prospective barrister, mini-pupillages are essential as they also show a commitment to and understanding of the Bar.[23] Barrister Katie Ghose also says that 'mini-pupillages are one way of finding out from other women how their experiences helped them to forge a practice at the Bar'. Further, there are chambers which will only offer pupillage to those who have completed a mini-pupillage in those chambers, seeing it as an extended interview.

Pursuing pupillage

Until 1996, there was no system for pupillage applications. Each individual made applications direct to chambers, often without knowing whether that set had any vacancies, and regularly without receiving a response, even a negative reply. This system came under considerable scrutiny because of 'allegations of racism, sexism and the old school tie'.[24] In addition, as the number of students taking the BVC increased, but the number of pupillages remained largely static, chambers were coming under increasing pressure around the time of pupillage applications. As a result, in 1996 the PACH (Pupillage Applications Clearing House) scheme was established.[25]

Candidates make one application via the PACH office, specifying a maximum of 12 sets of chambers to which they wish to apply. PACH then administers both the applications and offers, setting a timetable within which the process is to be completed. Around two thirds of sets are PACH members, although debate continues as to its value and some chambers have withdrawn from the scheme.[26] Nonetheless, there are a number of advantages to the PACH system, which have resulted in calls for it to be made compulsory.[27] The system has the potential to ensure a fairer system for all, with less reliance on prior connections, and it enables the collation of statistics regarding the recruitment process.

Table 5.3 gives the statistics for preliminary PACH outcomes as at January 1997. Slightly later figures show that the number of applications had increased to 1,772 for 843 pupillages, meaning over half of those who applied would be unsuccessful. It can be seen from table 5.3 that ethnic minority students are proportionately less likely than white students to be offered pupillages. Law Society research found no evidence of discrimination on the grounds of sex or ethnicity in the allocation of pupillages, although it did find a bias in favour of Oxbridge law graduates and students who had completed the CPE.[28] However, bias in favour of CPE students may indicate an element of indirect discrimination as those students are more likely to come from more privileged backgrounds and are more likely to be white than those who take law degrees.[29]

Table 5.3: Preliminary PACH outcomes by ethnic origin and sex

	White men	White women	Ethnic minority women	Ethnic minority men	Totals
PACH applications	855 (50%)	523 (30%)	176 (10%)	173 (10%)	1727
Offers made	499 (52%)	340 (36%)	58 (6%)	57 (6%)	954
Acceptances	277 (53%)	170 (33%)	39 (7%)	35 (7%)	521

The fierce competition for pupillages, and later tenancies, is reflected in the statistics given in table 5.4 which show the number proceeding through each stage of the qualification process. Of the 1,000 plus students who completed the BVC in 1995–96, 800 had a first six months pupillage, 760 a second six months and only 600 gained a tenancy. Part of the problem is that students must start the BVC before they know the outcome of any pupillage applications. By the October/November of the BVC, students may know that their applications have been unsuccessful, but they now have incurred the expense of the BVC – often amounting to at least £10,000 when living accommodation is taken into account. For this reason, the PACH dates should be brought forward so that students can decide whether to commence the BVC in the full knowledge of the outcome of their pupillage applications. This would reduce the financial hardship caused to those who pay for the BVC and then have to change career when they cannot secure pupillage.

Nonetheless, it is recognised that were such a change to be made, it may favour students from universities where the Inns and chambers have long established networks, who may therefore be more likely to gain a pupillage.[30]

Table 5.4: Total number of barristers commencing 1st six, and 2nd six month's pupillage and tenancy as at February 1997

Year	1st six months	2nd six months	Tenancy
1988–89*	560	489	386
1989–90	713	540	442
1990–91	917	627	419
1991–92	741	702	546
1992–93	859	767	555
1993–94	787	767	543
1994–95	852	773	574
1995–96	804	760	606
1996–97**	376	59	197

* From 1 Oct. to 31 Sept. ** From 1 Oct. to 10 Feb.

The competition for pupillage is, therefore, intense, and any experience that can show a commitment to a set of chambers or to the Bar generally will be an advantage. Barrister Katie Ghose notes that, in this competitive climate, the experiences of pupils before coming to the Bar have never been more important. Katie's previous experiences as an MP's researcher and parliamentary officer for the National Association of Citizens Advice Bureaux have stood her in good stead as she embarks on her career as a barrister. But Katie emphasises that many experiences, whether law related or not, can show experience, and mark out a candidate from the crowd. Legal experience may be gained by working for a firm of solicitors, law centre, or citizens advice bureau, mooting or debating at university and at an Inn, and writing and researching for legal journals. Mini-pupillages and shadowing a judge are the best ways of discovering the type of work and chambers which interest you.

When Elizabeth Gow was a pupil at Iscoed Chambers in Swansea she was involved in a free representation scheme in Wales and gained excellent experiences working in tribunals. She suggests that such experience is invaluable preparation for the shocks that await the unsuspecting in the higher courts.[31] In London, such a scheme is known as the Free Representation Unit, and there are similar schemes run in most of the circuits. Katie Ghose also advises that pupillage interviews are 'partly a test of stamina and performance under pressure and women who can show that they are accustomed to responding thoughtfully to difficult questions in a stressful situation are bound to impress'.

Choosing chambers

In light of the figures in table 5.4, the most pressing concern for most putative barristers is simply to get a pupillage, without being too specific about where. Nonetheless, there are many factors to be taken into account in making decisions regarding applications and, hopefully, accepting offers. The fact that some chambers pay a salary to pupils, which may range from around £6,000 to £15,000, with the larger commercial sets offering the most money, may be important.[32] There are also a number of possibilities for pupillage outside London and it is worth noting that chambers in the provinces tend to only take on the same number of pupils as there will be tenancies. As well as opportunities in the provinces, it is possible to train as a barrister within a company or other organisation. Barristers working in such organisations are part of what is referred to as the employed bar, and there are around 2,500 employed barristers in England and Wales, of whom just under 1,000 are women. Vacancies can arise with the Government Legal Service, the Crown Prosecution Service or local government.

Clearly the area of law is going to be a crucial factor. Elizabeth Slade QC has suggested that a putative barrister should look very carefully at the area of work in which she is considering specialising, not just in terms of interest and content, but also in terms of lifestyle choices, particularly the amount of flexibility in working practices.[33] Elizabeth Slade sees criminal law practice as the least flexible because of its reliance on court work, with family work similarly court-based. Common law practice is a mix of court and paper work. At the more flexible end, Elizabeth Slade suggests, are the public and commercial areas of practice which rely heavily on advisory paperwork, and where the court work is not as common and the cases have a very high settlement rate. However, the choice of specialism is a very personal decision and there will be a whole host of factors involved. The personal testimonies of women barristers at the end of this and the next chapter show some of the driving forces behind the choice of specialisms and show that it is also possible to re-orientate one's practice over time.

Although the days of men-only chambers have gone, the number of women in chambers still varies greatly, and may give an indication of the culture of the set, and the attitudes of members towards women barristers. Information about numbers of women in chambers can be easily obtained, although information about the culture of a set is more difficult to come by. Information gleaned on mini-pupillages and from keeping up to date with news about the Bar can help. As the personal testimonies of women barristers at the end of this and the next chapter demonstrate, and which is considered in more detail in the next chapter, the attitude of clerks and fellow members of chambers can have a considerable impact on a woman's practice at the Bar.

THE BAR'S EQUALITY CODE AND PUPILS

In addition to the new procedures introduced for the processing of pupillage applications, in 1995 the Bar Council adopted the Equality Code for the Bar which provides comprehensive guidance on all aspects of good equal opportunities practice. The Equality Code was adopted in the wake of the publication of research commissioned by the Bar Council and the Lord Chancellor's Department, entitled *Without Prejudice?*[34] This research detailed the depth of discrimination against women at the Bar. In particular, the researchers found that, despite comparable educational qualifications, on average women had to make more applications and attend more interviews for pupillage than men, and that more women have to 'squat' before gaining a tenancy. Further, there was a clear difference in the treatment of women and men at interview, with women being asked questions relating to their private lives and future plans, and too much reliance being placed on 'gut' feelings and 'intuition' in the applications process, at the expense of objective criteria. The report concluded that 'one of the biggest areas of disadvantage was at the junior level, when being selected for pupillage and tenancy'.[35]

As a result, the Equality Code sets out specific guidelines which should be followed in the recruitment of pupils and tenants. For example, chambers must have a written policy for the selection of pupils which should set out the job-specific criteria for the pupillages. This ensures that the criteria used for selecting pupils are the same for each candidate and that they relate to the position, otherwise 'the opportunity for unwitting bias and stereotyping and therefore unlawful discrimination is greatly increased'. No pupils should be accepted in chambers who have not come through the chambers' selection procedures, and no decision about any applicant should be made by one individual. The purpose of these recommendations is to avoid the 'old boy's network' which tends to indirectly discriminate against women and ethnic minority candidates. Chambers are recommended to use an application form, and not a CV, in order to ensure comparability of information about the candidates, and monitoring data on candidates should be maintained. This information will enable Chambers to see whether there are any obvious differences in the success rates of candidates of different backgrounds and sexes, which might merit further consideration.

At interview, the questions to be asked should be agreed in advance and the same ones asked of all candidates, with questions relating to personal relationships and family members being avoided. Further, the composition of the interview panel should include as diverse a group of members of chambers as possible, for example men and women of different levels of seniority and from different social and ethnic backgrounds. This is important because, as the Code notes, the 'composition of the panel may influence the applicants' perception of the interview process and hence their performance'. It is also to be preferred that members of chambers engaged in interviewing and selection undertake appropriate training.

The Equality Code applies not just to the applications process, but to all activities in chambers which might have a discriminatory effect on women. For example, in view of evidence that women are channelled into certain areas of work, it is recommended that the allocation of work to pupils is monitored every two months to ensure that there are no differences between pupils for which no account can be made (see following chapter). The monitoring required by the Code relates to the type of work, the fees paid and the reason why the work was given to the individual barrister. Similarly, pupilmistresses and pupilmasters are instructed to ask their pupils at regular intervals whether they feel that the distribution of work has been fair and whether pupils are satisfied that they have been given the opportunity to experience the full range of training available. The Code also deals with other forms of discrimination, such as sexual harassment, which will be considered in greater detail in the following chapter.

TOWARDS TENANCY

After pupillage, a pupil will need to secure a 'tenancy' in chambers, and many of the factors regarding pupillage choices will apply equally to applications for tenancy. A tenant is a full member of chambers, responsible for varying amounts of chambers' fees and rents. Being a junior tenant does bring a flexibility and control over workload that does not come with many other jobs. Self-employment means that it is up to the barrister to determine when she wishes to work, and where. The reality of the early years of practice, however, is long hours and very hard work. The barrister has to build up her practice and her relationships with solicitors, clerks and other members of chambers. Flexibility has positive and negative sides; it can mean irregular work, and being paid irregularly. It is also possible to move into the employed bar after pupillage.

Around 60 per cent of those who get pupillages go on to obtain tenancies. If not successful immediately in gaining a tenancy, all is not lost. Many pupils continue a third set of six months' pupillage, which can lead to greater experience and the possibility of a tenancy. In addition, many pupils go on to 'squat' in chambers, that is, building a practice whilst trying to secure a tenancy. The provisions of the Bar's Equality Code apply equally to tenants and squatters. Thus, the recommendations relating to selection of pupils will apply equally to tenants, as do provisions regarding the monitoring of the allocation of work, and more general issues regarding the treatment of women in chambers. The following chapter will go on to consider in more depth the experiences of women at the Bar.

CONCLUSIONS

We have seen in this chapter that becoming a barrister is highly competitive and that there may be some struggles along the way. But it can be worth it. Barrister Elizabeth Woodcraft describes her job as 'wonderful'. Laura Cox QC says: 'Life at the Bar is incredibly exciting, challenging and satisfying. If it is what you really want to do, if you are a very able lawyer, and if you have a real flair for advocacy, you should go for it!' The testimonies of the women barristers which follow, and those at the end of the next chapter, demonstrate the wide variety of work, the difficulties, and ultimately the rewards of a career at the Bar.[36]

This chapter has also considered the evidence that access to the Bar is not as open as it could be. It is promising that changes are being made and the collation of comprehensive statistical data would represent a further positive step. We will see in the following chapter that discrimination is a continuing problem at the Bar, and the first step towards progress is to open access. As Cherie Booth QC has argued: 'It is only by ensuring the best gain access to the profession and are promoted within it that we can be sure that the interests of justice are being served.'[37]

PERSONAL TESTIMONIES OF WOMEN BARRISTERS

Laura Cox QC, Cloisters

Laura was one of the first women to become a head of chambers. She has been at the forefront of action to improve equality of opportunity at the Bar and has acted in many of the most important cases in recent years which improve women's employment conditions. She discusses below her experiences as a pupil and in practice, and how the Bar is changing and becoming more welcoming to women ...

'My background was most definitely non-legal. There has never been, so far as I am aware, any lawyer in either side of the family. My father was a printer, running his own business, and my mother was a piano teacher. I went to a girls' grammar school in my home town and, from there, to London University (Queen Mary College) to read law. After the LLB, I went to LSE to do the LLM and studied trade union and labour relations law. This was in 1974 – old Labour was in power and the Sex Discrimination Act was just about to enter the statute books!

Women's rights at work, and generally, were just dawning in this country. The Equal Opportunities Commission (EOC) remind me that it was still common in those days to see advertisements for 'attractive girl Fridays' to work as receptionists, or for 'bright young men' to be lawyers' clerks. In 1974 a woman still had to leave the Foreign Office when she got married! Women at the Bar were still a rarity.

I don't really know when I first thought of law as a career. I remember always loving debate. My father used to take me to the old Quarter Sessions and I was attracted by the idea of advocacy from an early age. By A levels I was actively interested in civil liberties and human rights issues. At university this interest developed and I remember deciding, with a great deal of support and encouragement from the tutors at QMC and LSE, to have a shot at the Bar. At QMC, I won a mooting competition (the final was judged by Lord Denning) and the experience convinced me of the enormous value there was in seeking to persuade an independent tribunal to a decision in your client's favour, by a combination of legal knowledge and advocacy skills.

After Bar exams, I was very fortunate in obtaining a pupillage in Cloisters and, without doubt, in having Stephen Sedley, now Mr Justice Sedley, as my pupilmaster. I can truly say that my pupillage was the most wonderful experience; and I was so lucky to be offered a tenancy in Cloisters at the start of 1977. By the time I was a pupil (1976), Stephen was being instructed in some of the early sex discrimination cases. He was also instructed in personal injury cases, principally industrial accidents and industrial diseases, and later on in medical negligence work. I remember thinking that this was the kind of work I would really like to do. Stephen was undoubtedly an enormous inspiration. His brilliance as an advocate and a lawyer and his kind, unassuming nature had such an impact on me. At this time Mike Mansfield QC (though not then a silk) was a member of chambers too; and I remember being totally bowled over by his forensic skills and, especially, by his extremely thorough approach to case preparation.

Cloisters chambers has always been a mixed civil liberties set and there is no doubt that my forensic skills, such as they are, were formed and developed in my early years by the daily, relentless diet of Magistrates' Court criminal work. Liaising with court staff, cross-examining police officers, preparing complex legal submissions, often at very short notice, and gaining the solicitor's and the lay client's confidence were all skills you had to acquire very quickly if you wanted to do well. Later on, after I moved to a wholly civil practice, I continued to benefit from contact with chambers colleagues working in the criminal justice system. Such cross-fertilisation of ideas and approaches to a case, in these days of increased specialisation, has always been one of the main advantages of working in a 'multi-disciplinary' set, especially in the civil liberties field.

Undoubtedly, in the chambers I was in, I was fortunate in never having to deal with the sort of institutionalised discrimination experienced by many women at the Bar. The kind of work we did and the personalities of members of chambers created an atmosphere of mutual respect. If you worked hard and were good at your work, you could blossom. Of course having children (I have three sons) meant time out and a temporary drop in work (and income!). But again, in Cloisters, maternity leave was never an issue. It was taken for granted that I could have as much time off as I wanted and have financial assistance when I returned, to help me back into the fold. We were one of the first chambers to have a maternity and paternity leave policy in place.

After returning from my first period of maternity leave in 1980-81, I started to do more employment, discrimination and personal injury work, with a lot of trade union and EOC cases. This built steadily over the years and I also developed a practice in medical negligence so that, by the time I took silk in 1994, my practice was exactly how I had always wanted it to be.

I have also been fortunate to have the support of my husband over the years. Much of my success has been due to David, who has understood (though sometimes resented) the pressures and who has backed me all the way. I am well aware though both that I have been extremely fortunate and that, for many other women, the picture has not been so rosy. There has been real discrimination against women at the Bar over the years, in selection for pupillage and tenancy and then in access to work, pay and career development.

Without a doubt, however, there has been a sea change at the Bar over the last ten years. It is a real sign of the profession's genuine commitment to equal opportunities that the Bar has produced such a radical Equality Code for the profession and that it has taken, and is still taking, steps to ensure that it is understood and implemented. Statistics show that men and women are now being called to the Bar, and obtaining pupillage and then tenancy, in almost equal numbers. This equality at point of entry, together with a careful and constant watch over good equal opportunities practices in chambers, is bound to have an effect in the numbers of women succeeding at the Bar and eventually gaining silk or judicial appointment.

There has never been a better time for a women to come to the Bar. Of course the competition is fierce, both in finding a pupillage or tenancy and then in forging the practice you want. But there are two excellent equal opportunities officers employed by the Bar Council, who give first-line advice and information, together with a variety of networking groups and organisations to turn to for help and advice.

Life at the Bar is incredibly exciting, challenging and satisfying. Of course it means sacrifices. When you're in court and responsible for the conduct of a case it demands your attention unceasingly. Family and/or social and personal life will suffer for a while but, when the immediate pressure is over, insist on taking some time off to make up for it. Encourage your colleagues to do this too. Men, traditionally, have been reluctant to take time off the treadmill. In many quarters if you are not in chambers until 11pm each night and every weekend you're viewed as someone unlikely to succeed. The mould has to be broken (and not only, I may add, at the Bar!). This is not just a women's issue. We should all take responsibility for ensuring a better quality of life for ourselves and for our partners and children. There is, actually, more to life than the Bar, love it as I do.'

Dame Barbara Mills QC, Director of Public Prosecutions

Many people tried to dissuade Dame Barbara from becoming a barrister. She persevered and led a distinguished career in private practice before becoming the Director of the Serious Fraud Office and most recently the Director of Public Prosecutions. She describes below the development of her career and how she sees the future for women lawyers ...

'I was the first person in my family who ever wanted to be a lawyer. I was educated at a girls' school which puts much emphasis on academic achievement, including attention to detail and professionalism. My principal interests at school were in history and English literature, and I gradually became attracted towards the legal side of both these topics. By the time I was about 16, I wanted to become a barrister. Not surprisingly, in view of the very few women barristers who were then in practice, many people tried to dissuade me from what seemed to be an unrealistic objective. I persevered and was eventually encouraged by a barrister who told me that it was perfectly possible if I really wanted to do it. I never regretted my decision to become a barrister, even though from time to time it seemed to be against some serious odds.

I read law at Oxford and then passed my Bar exams. By the time I became a pupil in chambers I was married and had two children, and I have to confess that this was not much of an asset! In the early 1960s, a woman with or without children was not viewed as a long term investment for chambers. I hope that I did something to change that perspective. In due course, I found chambers who were sufficiently liberal to accept me, and I spent the rest of my career in practice at the Bar in those chambers. I am always grateful to them for having looked beyond the immediate and obvious problems to a longer term future. The chambers specialised in criminal law and so I spent my working life at the Bar both prosecuting and defending.

Inevitably, I started with small cases in magistrates' courts, but gradually my practice built up until I became Treasury Counsel at the Old Bailey, handling some of the most difficult and sensational criminal trials. In due course, I took Silk, and also became a Recorder, sitting as a judge. In 1990, my career changed dramatically. I became the Director of the Serious Fraud Office, and subsequently the Director of Public Prosecutions. The CPS is a Government department with approximately 6,000 employees, many of whom are lawyers and about half the lawyers are women.

My belief is that the future prospects for women in law are extremely promising. Many of our best new recruits are women. Organisations, including the Bar, are beginning to recognise that because so many women are practising in law, 'family-friendly' policies have to be adopted. At the Crown Prosecution Service we encourage women – and indeed men – to have career breaks to deal with family or other commitments and then to return to us. I am convinced that this is the right way for the future. Provided that these family-friendly policies are adopted, women can then have, as I have been fortunate to have, the great satisfaction of an interesting, important and rewarding career, combined with a happy family life.'

Elizabeth Woodcraft, Tooks Court Chambers

Elizabeth is a barrister committed to using her skills to improve the lives of others and to help bring about changes in the law and society. She has acted in many high profile cases and has worked in many campaigns to improve women's rights. Her practice is now mainly in the area of family law and she describes below her path to becoming a barrister, the work she is involved in, and talks about being a feminist barrister ...

'In 1976 I was working as a National Co-ordinator for the Women's Aid Federation, the organisation providing refuges for women escaping domestic violence. I worked closely with Jo Richardson MP on her Domestic Violence and Matrimonial Proceedings Bill, which was a major breakthrough in legislation for women by permitting the police to become involved in the civil world of injunctions. I so much enjoyed this part of the work that instead of returning to teaching I decided to read for the Bar. I was called in 1980 and in 1981 I joined Wellington Street Chambers, a radical set, famous for its fee sharing.

I grew up on a council estate in Essex, the middle daughter of a trade union official father and housewife mother. My family voted Labour and I was going on Aldermaston marches with my mum from the age of twelve. I was and always have been proud of my family's socialist politics. Coming to the Bar as a working class woman did not worry me since I wanted to do political work and thought, with some justification, that barristers already doing political work would not consider my class background a problem.

When I started practising, I worked in the areas of both crime and family. In the area of crime I represented Greenham women peace protesters, miners in the 1984 miners' strike, anti-apartheid protestors and, interestingly, women accused of beating their husbands. But eventually I came to find crime an inhospitable area of law for me as a woman. I found the courts harsh and unfriendly and the prospect of the cases I would be expected to do as a senior woman defence barrister (rape and sexual abuse of children) entirely unacceptable.

Family law represented everything I had come to the Bar to work for. The family is meant to be a haven of love and support for all members of society and is central to the lives of most women. However, it is where they suffer violence, abuse and oppression. Family law, like family life, is practised in private which means no-one knows what goes on. Family law is often seen as the natural home or ghetto for women barristers but I made a positive, political choice to do this type of work.

I do residence and contact cases involving domestic violence and sexual abuse of children. I represent lesbian mothers and gay fathers. Public law cases feature in my practice, where local authorities are making applications for care and supervision orders, and where often, women who are poor, have had little education and have had bad relationships, are put under pressure to be what the ideal mother is meant to be. I also work in the Civil Courts and appear before the Criminal Injuries Compensation Board (CICB), representing victims of rape and sexual and childhood abuse who are seeking compensation. In 1993 I represented 60 families at a public inquiry into how a male student nursery teacher had sexually abused their children.

These cases are intrinsically worthwhile and interesting, but I hope that I bring a slightly different perspective and ask questions which might not otherwise be asked. Individual cases don't necessarily change the world (although some can make a difference), but the right case going to the Court of Appeal, where the matter is heard in public, can raise an issue, start a debate and even make things better for all women.

Being a barrister is a wonderful job. It means that you can make yourself available to other organisations to be part of a wider process of changing things. I am a member of Rights of Women (ROW), an organisation of women working on legal issues affecting women. ROW organises conferences, undertakes policy work, writes books and pamphlets and carries out research. From 1991-1995 I was the Chair.

In 1985 I was part of a team of women commissioned by Liberty to investigate the issue of strip searching in Armagh jail, in Northern Ireland and in 1994 I was an observer of voter education in the run up to the elections in South Africa. More recently I have spoken on being a feminist practising Family Law, as part of a Gender Awareness Training course for members of the Indian Judiciary. I considered ruefully the impossibility of something similar happening in India for the British judiciary. I have also been a presenter of TV programmes on enforced caesarean sections, ethics committees and drugs trials, and surrogacy.

When I consider the future, the issues for me are whether to remain as a senior junior, or apply for silk, or even consider being a judge. Being a judge would mean I would not be dealing with the kind of cases I really want to do, except from time to time, and even then I would be constrained as to how

I dealt with those cases. This means to a certain extent weighing up personal ambition against political efficacy. Many of the cases which I do only arise at District Judge or County Court level, the kind of cases a QC would not do. On the other hand, to continue to be a respected member of the Bar it is expected that after a certain number of years you will move on to higher things. I think it is also true that we all want to see women getting ahead. I wouldn't of course be doing it for the status or the money. And if I ever become a QC that will be my story and I shall stick to it!'

NOTES

1   Contact details can be found in appendix five.
2   Michael Shiner and Tim Newburn, *Entry into the Legal Professions – The Law Student Cohort Study Year 3*, Law Society: London, 1995, p 95.
3   All statistics referred to in this chapter, including tables 5.1–5.4, were provided to the author by the Bar Council, unless otherwise stated.
4   Supra note 2, p 100.
5   One study found that women students earned on average £3.80 an hour, with men earning £4.55: *Times Higher Educational Supplement*, 25 July 1997.
6   Addresses can be found in appendix five.
7   See *Support for Students – a guide to grants, loans and fees in higher education 1998–9*, free from the Department for Education and Employment on (tel) 0800 731 9133. Many banks also offer career development loans and funding may also be available from some charities or colleges.
8   For further information about the AWB, see appendices five and six.
9   Quoted in *The Times*, 6 January 1998.
10   Colleen Graffy, 'Coming to terms with keeping terms', *Counsel*, March/April 1997, p 11.
11   Lord Justice Staughton quoted ibid.
12   *Times Higher Educational Supplement*, 13 December 1996.
13   Helena Kennedy, *Eve was framed – Women and British Justice*, Chatto & Windus: London, 1992, p 39.
14   Barbara Hewson, 'You've a long way to go, baby ...', 146 (1996) *New Law Journal* 565–566, p 565.
15   Waterlow's, *Solicitors and Barristers Directory*, 1996.
16   See chapter six under the heading 'No room at the Inn?'.
17   A drop from 87 per cent the year before: Inns of Court School of Law Press Release, 9 August 1996.
18   Patricia Wynn Davies, 'Prospects for black barristers get worse', *The Independent*, 13 March 1997.
19   Final Report of the Committee of Inquiry into Equal Opportunities on the Bar Vocational Course (the Barrow Report), 1994.
20   Ibid. For further information about the SBL, see appendices five and six.
21   See chapter one under the heading 'Degrees of success' and chapter three under the heading 'Passing the LPC'.
22   *The Lawyer*, 5 August 1997.

23   An application should be made to the person in charge of mini-pupillages in each set of chambers. Because of the large number of applications, some chambers may give priority to candidates who are in their final year at university or doing the CPE. Information about mini-pupillages can be found in the *Chambers Pupillage and Awards Handbook* which is available free from the Bar Council.

24   Jonathan Kirk and Andrzej Bojarski, 'A Modern Route to the Bar', *Counsel*, May/June 1997, pp 9–11, p 9.

25   Details of the pupillages available and of the PACH scheme can be found in *Chambers Pupillage and Awards Handbook*, supra note 23.

26   As reported in *The Lawyer*, 1 July 1997.

27   Supra noted 24.

28   Supra note 2, pp 126–7 and Michael Shiner, *Entry into the Legal Professions – the Law Student Cohort Study Year 4*, Law Society: London, 1997, pp 125–6.

29   Supra note 2, pp 126–7. See also chapter one under the heading 'Common Professional Examination'.

30   See Nigel Bastin, 'New pupils for a new century', *Counsel*, April 1998, pp 30–31.

31   Elizabeth Gow, 'Free representation', *Counsel*, November/December 1996, p 20.

32   About one half of the pupillages obtained in 1995–96 were funded.

33   Elizabeth Slade QC, Conference Address, 'Law: A Flexible Career?', St Hilda's College, Oxford, 7 March 1998.

34   TMS Consultants, *Without Prejudice? Sex Equality at the Bar and in the Judiciary*, Bar Council and LCD, 1992.

35   Ibid, p iii.

36   See further Fiona Boyle, *Legal Careers*, Fourth Estate and Guardian Careers Guides, 1997.

37   Cherie Booth, 'Equality: The Bar in the 21st Century', 2 (1998) *Inter Alia* 3–4, p 4.

Chapter 6
Without prejudice?
Women at the Bar

'Some years ago it was claimed that women were discriminated against in seeking to become and remain barristers. Whatever the truth of this accusation, it is clearly now the case that women at the Bar are competing on merit with their male colleagues.'

Michael Wright, Attorney General and Chairman of the Bar, 1983[1]

'The results of the survey showed substantial evidence of early and continuing unequal treatment of the sexes at many levels in the profession. ... The report concludes that gender discrimination appears to be institutionally present within the Bar.'

Without Prejudice? Sex Equality at the Bar and in the Judiciary, 1992[2]

In early 1998, the BBC turned down a new series of Rumpole of the Bailey, concerned that the 'sexist' views of Rumpole were out of date, as were those of the feminist woman barrister in the programme.[3] The BBC considered that 'the feminist politics of the last Rumpole offering were dated and old-fashioned'. The unfortunate reality is that the sexist views of Rumpole do still reflect commonly held assumptions at the Bar, and that the 'feminist' politics of women barristers is needed now as much as twenty years ago. Indeed, it is only in the 1990s that a real debate on the situation of women at the Bar has been generated within the profession, with action being taken as a result. Accordingly, whereas the BBC were accused of 'covering up the nasty underside of life for women' at the Bar, this chapter will attempt to shine some light on the reality of practice for women barristers.

This chapter first draws on the statistical evidence regarding the number and status of women at the Bar and goes on to consider some of the problems and difficulties facing women in both developing and sustaining their practice. It then examines the measures which are being taken to alleviate potentially discriminatory practices. The chapter concludes with the personal testimonies of three women barristers which, together with those at the end of chapter five, exemplify and expand on many of the issues considered in this chapter.

ELUSIVE EQUALITY

Calling women

The number of women being called to the Bar has increased dramatically from a paltry eight per cent in 1970 to 43 per cent in 1997 (see table 6.1). These statistics do not, however, paint the whole picture. Being called to the Bar is not synonymous with gaining a tenancy in chambers and practising, just as qualifying as a solicitor does not guarantee employment as a solicitor. Thus, many who are called to the Bar will not go on to practice, either through choice or the inability to gain a tenancy. We know that in 1995–96, 606 tenancies were filled (see table 5.4), but this statistical data does not break down the numbers in terms of sex and ethnicity. Thus, it is not clear whether women and men are awarded tenancies in proportionate numbers.

Table 6.1: Number of women and men called to the Bar 1970–1995[4]

Year	Men	Women	% Women	Total
1970	858	77	8%	935
1975	890	112	11%	1,002
1980	617	245	28%	862
1985	663	282	30%	945
1990	507	339	40%	846
1995	599	507	46%	1,106
1996	1,034	651	39%	1,685
1997	910	698	43%	1,608

There is however some evidence of potentially discriminatory practices when women are applying for tenancies. Research commissioned by the Lord Chancellor's Department (LCD) and the Bar Council in 1992, resulting in the welcome publication of the report *Without Prejudice? Sex Equality at the Bar and in the Judiciary* (hereafter the *Without Prejudice?* research) showed that 39 per cent of women were asked in tenancy interviews about their marriage and children, compared with only 15 per cent of men. As the researchers noted, the asking of such questions is 'powerfully indicative of discriminatory practices and attitudes'.[5] There is much anecdotal evidence that such practices continue. Barrister Katie Ghose records that 'women who have spent years juggling commitments to family and work are still asked who will take the kids to school if she has to be in chambers by

8.30am'. At the 1998 Woman Lawyer conference one woman pupil reported that she had been asked at five out of six tenancy interviews about her childcare arrangements.[6] The statistical information which is available on 'squatting', that is, practising from chambers without being a full tenant, also reveals differences between the sexes. The *Without Prejudice?* research found that 36 per cent of women respondents had squatted, compared with only 18 per cent of men. In addition, women, on average, have to squat for a much longer period before gaining a tenancy than men.

Some information is also available on the rate at which barristers leave independent practice. 'Drop-out' rates for all barristers are extremely high, but the *Without Prejudice?* research found that the survival rate for women barristers was less than for men. Since 1970, the survival rate for men was 34 per cent, compared with 29 per cent for women. Indications are that most barristers leaving independent practice do so within two years of being called and the principal reasons for leaving are the failure to secure a tenancy and low earnings, although more women than men leave to care for dependents. In total, there are 9,369 barristers in independent practice, of whom one quarter (24 per cent) are women.

Queen's Counsel

The conferment of the title Queen's Council (QC) is a mark of distinction, signalling the achievement of senior status in the profession, with around ten per cent of the practising Bar having taken silk. To be eligible, applicants must be advocates (barristers or solicitors) who have full rights of audience. It is usual that the applicant will have been in practice for at least ten years, though in reality applicants will have between 15 and 20 years of experience. Being a QC can be an asset when pursuing a judicial career, and is certainly a passport to more high-profile work and greater fees.

At the Bar Council's Annual Conference in 1996, Sir Christopher Rose, a senior Appeal Court judge, said that he knew of examples where women had been appointed QC only because of their sex, and where a man of comparable ability would not have been appointed.[7] The figures speak for themselves. As of April 1998, there are just under 1,000 QCs of whom 73 are women, making seven per cent. 13 QCs are from the ethnic minorities, but no statistics are available as to whether they are women or men.

Table 6.2 shows that the number of women appointed to silk each year is very low, with 1998 being the first year to get into double figures. The figures for 1996 and 1997 actually show the smallest number of women appointed for some years, even though more women are applying (although the figures are small enough that a few applications either way makes a difference). 1998 did see an improvement, although it appears the more remarkable only because of the figures in 1996–97. In addition, 1998 saw the smallest number of silk appointments for many years, meaning that the percentage representation of women appears considerably higher than in earlier years. Nonetheless, the success rate for women applicants has been slightly higher than for men.

Table 6.2: Women and men QCs: applications and awards 1991–1998[8]

	1991	1992	1993	1994	1995	1996	1997	1998
Women applicants	24(6%)	34(8%)	37(8%)	43(8%)	42(9%)	40(8%)	41(8%)	46(9%)
Men applicants	386	386	435	496	450	448	459	465
Total	410	420	472	539	492	488	500	511
Women silks	6(8%)	7(10%)	6(9%)	9(12%)	8(11%)	4(6%)	5(7%)	10(17%)
Men silks	67	62	64	68	63	62	63	50
Total	73	69	70	77	71	66	68	60
% women applicants awarded silk	25	21	16	21	19	10	12	22
% men applicants awarded silk	17	16	15	14	14	14	14	11

It is often suggested that the low number of women QCs results from the historical fact that women have only been admitted to the Bar in sufficient numbers in recent years. This explanation is not, however, borne out by the facts. In 1996, of those juniors with over 15 years' call, 12 per cent were women (1 in 8).[9] Yet, at the same time, women made up only 1 in 15 of the QCs in private practice.[10] In 1998, 14 per cent of barristers with over 15 year's call are women.[11] Accordingly, there are many senior women barristers eligible to take silk. Therefore, other reasons must be investigated as to why there are so few women QCs.

Table 6.3: Ethnic minority QCs: applications and awards 1992–1998[12]

	1992	1993	1994	1995	1996	1997	1998
Applicants	14(3.3%)	14(3%)	11(2.3%)	12(2.4%)	14(2.9%)	12(2.4%)	18(3.5%)
Awards	1(1.4%)	1(1.4%)	1(1.3%)	1(1.4%)	1(1.5%)	1(1.5%)	4(6.7%)
% applicants awarded QC	7%	7%	9%	8%	7%	8%	22%

Figures for QCs of ethnic minority background are shown in table 6.3. There are no statistics for ethnic minority women. In 1992 only 14 out of 420, that is 3.3 per cent, of applicants for appointment as QC were from ethnic minorities, and since 1992 the proportion of ethnic minority applicants actually fell to 2.4 per cent in 1997, although in 1998 it returned to 1992 levels. In addition, 1998 saw four ethnic minority applicants appointed as QC, an improvement from the usual solitary success. Although only one per cent of barristers with 15 years of experience are from ethnic minorities, it is encouraging to see that of those with five to ten years of experience, eight per cent are from ethnic minorities.[13]

Representing women

In addition to the numerical superiority of male barristers in practice, barrister Barbara Hewson has written about the lack of women in any of the governing positions of the Bar.[14] Thus, there are few women leaders of Circuits, of the Bar Council's officers, of the Inn's representatives on the Bar Council, of the Benchers of the Inns, and of the Inns' Council. These roles are important as they are the decision-making and policy-directing bodies of the Bar. In 1994, the Bar Council turned down an application by the Association of Women Barristers (AWB) to have a designated seat on the Bar Council to ensure a greater representation of women.[15] As with a similar request to the Law Society, it was rejected, despite the fact that the 1979 Royal Commission on Legal Services had recommended that such a step be taken.[16]

It might be said that the time has been reached when the Bar Council can claim an element of meritocracy because in 1998 it elected its first woman chair of the Bar Council. Heather Hallet QC's appointment was announced in *The Times* with the headline 'Working mum reaches the top', with Ms Hallet being described as 'blonde and bubbly'.[17] It is certainly clear that these are not descriptions which would greet a male incumbent of the post, although doubtless many previous Bar Chairmen have had children. Nonetheless, her appointment is symbolic of the changes which are taking place at the Bar, although this process of change is far from complete.

PROBLEMS IN PRACTICE

Why is it that women appear to have greater difficulties than men in gaining tenancies, continuing their practice and taking silk? The *Without Prejudice?* research found strong feelings expressed in favour of the status quo at the Bar. Comments such as 'that's the way things are ... individuals need to accept it' and that 'it's tough for everyone, not just women' were commonly made. It was also a generally expressed view that women will succeed according to their ability. In a similar vein, black barrister John Taylor denounced the Bar's Equality Code (see later) as being unnecessary because, apparently, 'ability, like cream, floats to the top'.[18]

We saw in chapter four that such views can be characterised as a 'human capital' perspective on the role and situation of women.[19] In other words, it is said that women will succeed according to the amount of effort and commitment - 'human capital' - that they put into their career, and that accordingly women's lack of success is due to their own lack of commitment, enterprise or ability. An added variant of this perspective is the seldom voiced view that women who have children are necessarily demonstrating reduced commitment to the workplace, and, due to certain innate feminine qualities, women will naturally gravitate to certain types and levels of work. The 'human capital' explanation tends to dominate discussion of the role and situation of women in most workplaces and professions. For example, Hazel Williamson QC told the Woman Lawyer conference in 1995 that women 'can no longer blame "discrimination" for [their] lack of success'.[20] The implication is that any 'blame' must be attached to the women themselves.

It is no doubt the case that there are women, as there are men, who do not seek seniority in their chosen profession, or who do not invest time and effort in their careers, and such individuals would indeed only have themselves to 'blame' for any lack of 'success'. However, what is equally clear is that there are structural disadvantages, or, as the *Without Prejudice?* research termed it, institutional discrimination, which has a deleterious affect on women's careers despite their best efforts. Any such account of women at the Bar, therefore, must take into account gender as a basis for describing and investigating the careers of women barristers.

'Invisible pregnancies and self-raising families'

Baroness Helena Kennedy QC commented in 1978 that 'central to the whole problem of discrimination [at the Bar] is the issue of childbearing'.[21] Not much has changed. In 1990 a woman barrister wrote to *The Independent* suggesting that at the Bar 'women are still expected to have invisible pregnancies and self-raising families'.[22] Thus, one of the principal problems for women at the Bar, and in most other workplaces, simply results from the biological capacity to give birth. Not only does the very fact of pregnancy cause disquiet in many male-dominated workplaces, especially the more visible the pregnancy, but there is an assumption that women should be the principal childcarers and that childbirth necessarily signals a reduced commitment to work.

Although barristers are self-employed, and theoretically their working arrangements should permit greater flexibility, often financial constraints and responsibilities to colleagues and chambers seriously inhibit taking a break from work.[23] Being self-employed means not earning during a period of maternity leave (or not earning for a period of time after a return to work), thus increasing the pressure to return to work as early as possible. In addition, the longer a barrister is unable to work, the longer existing clients have to find other barristers to do their work.

On the subject of combining work and a family, barrister Geraldine Andrews says that she has 'complete admiration' for those who have managed to fit in the demands of a family with work. Geraldine has remained single and regards this 'as part of the price which I have had to pay for climbing the ladder towards the top of my profession'. It appears that Geraldine may not be alone. The *Without Prejudice?* research found that 61 per cent of women respondents to the survey had children, compared with 85 per cent of men. This is a message which some women law students are picking up. In a recent survey of women law students, one barrister-to-be responded that in ten year's time she saw herself as 'a practising barrister with no children because of the existing lack of sympathy for women barristers who want to also have a family'.[24]

Indeed, one Head of Chambers remarked in the *Without Prejudice?* research that 'maternity is seen by some chambers as a likely problem for them to be avoided by not having too many women', and one woman barrister stated that 'there is a clear policy in chambers that women are not to have children'.[25] The majority of women taking maternity leave return to chambers within six months, many within a few weeks. Ruth Deech, Principal of St Anne's College, Oxford, suggests that the Bar is a 'difficult profession to leave for a few years while raising a family' and she would not recommend taking time out. The *Without Prejudice?* research found little evidence of women returning to the Bar after a lengthy career break and that if they did return to practice after such a break, they were more likely to return to the employed Bar or the solicitors' profession. Indeed, this is the experience borne out by District Judge Gillian Marley. After eight years of practice at the Bar, Judge Marley had two children within 14 months of each other and although she had intended to continue her practice on a part-time basis, she found it too difficult to do so. She took a career break for five years to look after her children, later re-entering the profession by re-qualifying as a solicitor.

The Bar Council adopted a maternity leave policy in 1992 which provides that a woman's seat in chambers should remain open for up to one year while she takes maternity leave. The Bar Council also recommends that chambers should offer a period of three months' maternity leave, free of rent and chambers expenses. The maternity policy, which is now incorporated into the Equality Code, also informs chambers that many barristers may wish to return to work much earlier than the one year entitlement, and that many women may wish to carry out drafting and advisory work during maternity leave. However, the Code has not been implemented by all chambers and a survey by the Bar Council found that of those chambers which responded (between one quarter and one third), only 65 per cent had a maternity policy.[26]

Of course, many women do manage to forge a successful career at the Bar and have a family. Dame Barbara Mills QC, Director of Public Prosecutions, had two children by the time she became a pupil in chambers and says that she has to 'confess that this was not much of an asset'. 'I have been fortunate', says Dame Barbara, 'in being able to combine a most interesting career, together with a happy family life'. Laura Cox is a highly successful QC and Head of Cloisters Chambers and has managed to combine this success with having three sons. Laura acknowledges the support she has had from her husband in this, and notes that other women have not been as fortunate as her in the supportive work environment that she has enjoyed at Cloisters.

Women's work

The Lord Chancellor recently acknowledged that 'too often in the past, women were channelled into what were thought of as "appropriate" areas of law, for example family and matrimonial work' and he recognised that many still believe this to be the case.[27] Women may be directed towards this work on the basis that this is 'women's work', and that areas of work such as crime or commerce 'are considered to require assertive or aggressive practitioners, and women are not thought to have the necessary qualities'.[28] But it is not just in respect of the area of work that women may be affected by stereotypical assumptions, but also in the nature of the work, in that women may often not be considered for more high profile cases or those that appear more complex.

These difficulties can arise because considerable amounts of work come into chambers on an 'unnamed' basis, that is no particular barrister is requested. When this happens, the power of clerks to distribute work is crucial. It is at this stage that women pupils or barristers may be disadvantaged by the stereotypical assumptions of clerks, in terms of not giving work to certain individuals and in directing only certain types of work to women and/ or ethnic minority barristers. Concern over the role of clerks has been expressed for many years, with the publication in 1993 of research detailing the discriminatory attitudes of many clerks to women and ethnic minority barristers causing particular controversy.[29] Indeed, by 1997, the Commission for Racial Equality (CRE) was sufficiently concerned about the attitude of clerks that it threatened an investigation into their practices.[30] These concerns were echoed in the *Without Prejudice?* report which found that just under one half of head of chambers had experience of clients expressing a preference about the gender of the barrister and that in many cases this preference was accommodated. In addition, the research found that fewer women (68 per cent) than men (89 per cent) practice in their original area of preference and that there exists a pattern of men working in commercial law and women working in family law and certain branches of criminal law.

Problems also arise from the assumptions of solicitors. Najma Khanzada is a criminal defence barrister and finds that in her work, clients – lay or professional (solicitors) – want to instruct a 'bruiser', and their first choice is not a young woman. Alternatively, clients sometimes wish to instruct a woman to defend alleged rapists and child abusers on the basis that the fact that a woman is defending them will help their defence. One of the reasons which led Elizabeth Woodcraft to move into family law was the 'entirely unacceptable' prospect of having to accept such instructions as a senior woman defence barrister.

The effect of such customs is that areas of practice arise which are seen as 'women's work', and areas of law or types of cases which are considered more appropriate for men. This also has a knock on effect in terms of the status accorded to certain areas and remuneration levels. The tendency is that women are remunerated less than men because their areas of practice, such as family law, are those which are less well paid than, for example, commercial law, and because women and ethnic minority barristers simply do not have the same 'purchase' as white men.[31] This is a particular problem as one of the factors considered in the appointment of QCs is the level of salary, founded on the rather superficial belief that a high salary indicates a successful practice.

One of the means by which these problems can at least be identified, and then corrected, is by monitoring the allocation of work in chambers. We noted in the previous chapter that the Equality Code recommends that the allocation of work to pupils and junior tenants, in particular, should be monitored, but the same principal applies to all members of chambers. Thus, work between members of chambers, unnamed work which comes into chambers, and fee levels should all be monitored in order that any discriminatory patterns may be traced. Without this knowledge, it is simply not possible for chambers to claim that they are committed to equal opportunities. Detailed guidance on methods of monitoring are given in the Equality Code and elsewhere.[32] The role of the clerk is also changing. Clerks have traditionally earned around ten per cent of barristers' earnings, but this percentage has been reducing over the years, and many chambers now employ salaried practice managers. Some practice managers are drawn from outside the law, and therefore are not imbued with the 'culture' of clerking which may be inimical to the interests of some barristers.

Sexual harassment

Sexual harassment encompasses a wide range of actions, from physical assault to demands for sexual favours in return for career advancement, to compromising suggestions or looks, jokes or abuse, to unnecessary and avoidable physical contact. Women pupils are in a particularly vulnerable position, but sexual harassment can affect any woman, in any area of work. The *Without Prejudice?* research reported that sexual harassment is not always recognised as such, as it is often thought by the perpetrators to be something of a joke, a bit of fun that everyone must accept and get used to. One woman QC responded during the research that 'the Bar is male-dominated and a woman will have a tough time if too sensitive to, for example, sex orientated jokes'. However, such a comment is borne of an ignorance of the effect which sexual harassment can have on an individual. Sexual harassment is an abuse of power designed to undermine the recipient and their professionalism, confidence and safety. As recognised by the law and by the Bar's Equality Code, the fact that one person may be able to ignore or deal comfortably with certain behaviour, does not mean that it is acceptable if directed at another.

A 1995 study found that 40 per cent of women respondents had encountered sexual harassment at the Bar.[33] Although the sample involved was small, the research is suggestive of some broader concerns and is supported by other studies. A report by an internal working party of the Bar Council on pupillage and training found alarming levels of harassment by pupilmasters, denouncing the conduct of some as 'disgraceful'.[34] Respondents to the *Without Prejudice?* research reported that it was the belittling 'witticisms' and the 'constant drip of irritating remarks' which were problematic.

Barrister Barbara Hewson caused a furore when she suggested that the level of sexual harassment at the Bar was 'unacceptably prevalent'.[35] An article which she wrote in the New Law Journal on this was reported in much of the national press, with a particular emphasis on the fact that she reported that she receives a 'trickle of complaints' each year about offers of pupillages for sex.[36] Barbara says that the reaction in some quarters was to downplay the problem and criticise her for ventilating it. 'It is, perhaps, a telling reminder', says Barbara, 'that male-dominated institutions find harassment a particularly difficult subject: it involves facing up to some uncomfortable realities about male power, which some people – women and men – find hard to deal with'.

What confounds the problem of sexual harassment is the understandable unwillingness on the part of those who have experienced harassment to come forward and make a complaint. If a woman does experience harassment there are a number of avenues of redress open to her, but some of these steps inevitably involve a considerable undertaking on the part of the complainant, and what most often characterises sexual harassment is the wish on the recipient's part not to go public.[37] As a last resort, the Equality Code states that the Chair of the Bar Council may request chambers to volunteer to act as 'safe havens' which will provide pupillage, or an opportunity to squat, to pupils or junior tenants who are unable to remain in their chambers as a result of harassment. This particular recommendation met with considerable publicity and a certain amount of unjustified ridicule in the legal and national press, but recent events have shown it to be of necessity and of practical application. This procedure was used for the first time in early 1998 when a woman pupil was able to move chambers, after a Bar disciplinary tribunal found a barrister and part-time judge guilty of persistent sexual harassment.[38] Of course, it is the woman who has been harassed who has to move chambers, almost as if she is the one being punished, but, nevertheless, these procedures do represent a step forward and a form of collective action by the Bar.

Chambers need to take steps to implement the provisions of the Equality Code and a model harassment policy for chambers has been drafted for this purpose.[39] Undoubtedly, the total elimination of sexual harassment is unlikely. What can be achieved, however, is the development of a culture and ethos in which women feel able to bring complaints, without committing professional suicide, and in which sexual harassment and the sexualisation of the workplace is not tolerated.

Ethnic minority women barristers: a colour Bar?

'No-one ever dreamed Lee Mitchell would be a barrister – except Lee herself. Black, working-class, a woman, she has little on her side except determination and intelligence.'[40]

This is how Nicola Williams introduces us to the protagonist in her novel *Without Prejudice*. Probably one of the few black woman barristers to feature in a novel, Lee Mitchell provides a role model for black women contemplating the Bar: a fictional role model being required as there are so few real ones. At the 1995 Woman Lawyer Conference, black barrister Anesta Weeks recounted how racist and sexist attitudes within the profession are burdens on her work which she resents. She also pointed out that, aside from the BVC, there are no statistics collated on ethnic minority women barristers and it is therefore very difficult to understand the true position, or monitor change. She talked about obtaining work as the battle of the three males: the client, the solicitor and the clerk. Black women suffer because the three males often have a pre-conceived negative attitude about an image before a jury trial: a black woman fighting their corner is not something to which they are accustomed.[41]

Najma Khanzada is a criminal defence barrister and describes the problem of race and gender discrimination at the Bar as arising from stereotypical attitudes. 'This discrimination is not life threatening, one's human rights or even basic employment rights are rarely violated', but nevertheless, the discrimination 'impinges on the way one does one's work and even the quality of work'. Women are a less marketable asset in crime and solicitors are less inclined to brief junior women practitioners for fear of disappointing the client. Najma has found that even after three years' practice 'some solicitors still feel it necessary to explain the simplest criminal procedure to me'. Najma notes that at one level the solution to the problem is to present oneself so that no client, no opponent and no judge is entitled to make prejudicial assumptions. 'But this solution denies the existence of the problem. The problem is that, however assertive, however technically competent you are, however bullish, you are perceived in a particular way. It is not you who has to change, it is those who perceive you who must.'

The Bar's Equality Code prohibits discrimination not just on the ground of sex, but also of race, as well as other forms of discrimination. Thus, all of the provisions relating to monitoring of work, harassment, selection for pupillages and tenancies apply equally in the case of race or ethnicity. However, steps also need to be taken by the Bar Council and the LCD in order to ensure that statistical data is collated which is broken down in terms of sex and ethnicity. The issue of ethnic representation in the profession was the subject of a conference in 1997, the Minority Lawyers Conference, which sought to address problems of the under-representation of ethnic minority lawyers throughout the legal professions. The CRE also launched in 1997 an initiative to raise the profile of discrimination against ethnic minority women in the population as a whole. The campaign, 'Visible Women' aims to raise awareness about the aspirations of ethnic minority women and to encourage employers and women's groups to take action to increase ethnic minority women's opportunities. In 1998, the Society of Black Lawyers (SBL) will be publishing an updated version of its directory of practising black lawyers – solicitors, barristers and other legal practitioners.[42] Encouragingly, therefore, positive action is being taken, although there remains a long way to go.

Taking silk

We saw above that the number of women taking silk, although increasing, remains very low, raising questions about why this might be the case. It can be argued that the principal factor is the appointments process which, as with judicial appointments, is based on what are known as 'secret soundings', that is consultations made by the LCD of senior members of the Bar and judiciary. The *Without Prejudice?* researchers described such a process as one of 'patronage': a system of appointments based on the 'old boys' network', with an in-built advantage for men which does not favour equal opportunities.[43]

Secret soundings form the backbone of the appointments 'process' for both the judiciary and silks, but their use for the latter is more inimical than for many appointments to the judiciary. We will see in the following chapter that in the appointments process for many judicial positions, lay people are involved at an early stage in the selection process, both in terms of initial assessment of applications and at the interview stage.[44] In contrast, the process for silks involves no lay element and no interview process and this system is therefore much more heavily reliant on the views of the judiciary and senior members of the profession.

In addition, the selection criteria are slightly different (although many would argue that the actual assessment of applicants against the criteria is improbable). In particular, the emphasis placed on earnings, albeit with the caveat that remuneration will not be a 'decisive' factor, may have negative effects on the applications of many women and practitioners from an ethnic minority background. As noted above, the *Without Prejudice?* research found that women barristers were generally paid substantially less than men. One of the other more worrisome criteria is that the barrister must be of the 'highest professional standing, having gained the respect of the Bench and the profession'.[45] It used to be the case that 'standing' in the profession was a criteria for the judiciary. Standing is a very subjective notion that is prone to stereotypical interpretation. For example, as we have seen throughout this book, women tend to be seen as less authoritative than men. This is due to historical and cultural factors such as women being more associated with irrationality, emotions, the home and childbearing, which compares unfavourably with the mythical image of the rational, objective, neutral man of law. Thus, 'standing' for women is going to be more difficult to achieve. Standing also equates to being 'known' within the profession, something which is much easier on the whole for men who have greater access to contacts and networks.

At the 1997 Woman Lawyer conference, Heather Hallet said that to become a QC one has to get known and that she took a deliberate decision to become involved in Bar politics for this very reason. She attributes this involvement to her own success, suggesting that one of the reasons why women may not succeed in reaching senior positions at the same rate as men is that: 'When women are not in court, they want to be at home, even working from home. They do not want to go to the Saturday conferences and stay late for Bar political meetings and other things that barristers do to get up the ladder, so they miss out.'[46] In many ways, of course, this simply underlines the fact that to get on, one has not just to be good, but to be known.

The 1994 Kalisher Report presented the Bar's proposals on reform of the silk system, but made no reference to the *Without Prejudice?* report. This omission was later explained as an 'oversight', but as Barbara Hewson has argued, it may have had something to do with the fact that the general conclusion of the Kalisher report was that the present structure of silk appointments should be retained.[47] The Kalisher Report included some examples of the comments that have been made by consultees in the process of secret soundings. For example: 'I am surprised to see that he is 48, he looks younger'; 'I don't know him personally but those I have consulted say that he is OK although not very special'; '[ ] QC says he is not silk material. I do not know him personally'.[48] Presumably, these comments were given as examples of the good practice of the soundings process. Although these are extracts, there appears to be no reference to any selection criteria, rather it appears that consultees rely on a 'gut' instinct and the views of their acquaintances who do not have to explain and justify their comments. In addition, comments are included from those not on the formal consultee list, and which would appear to amount to little more than gossip. Furthermore, it seems clear that judges do not keep contemporaneous records of how they consider barristers to have performed before them, and therefore the consultations rely on remembered perceptions which can quite easily reflect factors which have little to do with the competence of the individual.

This is an appointments system which is in need of immediate reform, especially in view of the links between taking silk and becoming a judge. There is little point in reforming the applications process for the judiciary, when a preliminary to many such appointments, taking silk, is based on an even more antiquated and potentially discriminatory process. An editorial in *The Lawyer* declared in 1997 that: 'the Lord Chancellor's Department is determined to carry on with a promotions system that is perceived to be so stacked against most barristers that they do not even apply, that alienates all but white men at the Bar and that is, quite simply, out of date'.[49] *The Lawyer* recommended a 'complete rethink' with independent reviewers looking at the system. Ruth Deech, Principal of St Anne's College, Oxford and chair of the Human Embryology and Fertilisation Authority, suggests that the main hurdle for women at the Bar is the 'lack of transparency in appointment to silk and to the bench. Where the promotions processes are invisible, then I fear that at the Bar, as well as in other spheres of life, women's merits might not be given full credit and that in borderline situations, the appointers prefer men.'

A number of suggested reforms will be considered in the next chapter, but as well as calls for reform, there are advocates of the abolition of silks. Barrister Barbara Hewson argues that although judges would loose their easily identifiable pool from which to draw judicial appointments, as this is a male pool, its abolition will not affect women.[50] Josephine Hayes, chair of the AWB, is similarly supportive of abolition. She notes that the rank of QC is a traditional Government seal of approval for advocates, but recognises that even if there was a need for senior recognition of advocates, one might have thought that the legal profession itself was up to the task of selection without Government assistance.

No room at the Inn?

The Inns of Court are highly influential bodies which are run by Benchers who tend to be either High Court judges or senior QCs. It was noted in the previous chapter that the number of women Benchers is extremely low, only six per cent, and that appointment is made by invitation.[51] For two years running, the Inns have been censured at the Bar's AGM for failing to introduce a comprehensive equal opportunities code.[52] The chair of the Bar's Race Discrimination Committee, Lincoln Crawford, has described the failure of the Inns as 'inexcusable'.[53] Twenty years ago, the Royal Commission on Legal Services recommended that the Inns elect more women as Benchers.[54] The fact that this recommendation has to be reiterated after such a long time is unfortunate, but necessary.

Barbara Hewson has argued that the lack of women Benchers is important as 'it is not unreasonable to suppose that your chances of judicial appointment, or silk, are improved immeasurably if you lunch with senior judges every day'. She further questions the need for four Inns of Court which spent one million pounds a year on catering and just over one and a half million on scholarships. Finally, she suggests that they no longer offer a collegiate role, and are extremely resistant to any questioning especially regarding equal opportunities.[55] Another critic of the Inns, Martin Bowley QC, has described the Inns as 'legal theme parks' inhabited by 'self-perpetuating, geriatric, oligarchies'.[56] It certainly appears clear that at a time when the Bar is trying to improve equality, the Inns are out of step.

THE EQUALITY CODE AND BEYOND

The Bar has taken considerable steps to try to stamp out discrimination within its ranks in recent years, what Laura Cox QC describes as a 'sea change', with the adoption of the Bar's Equality Code playing a central role. The drafting, consultation, review and adoption of the Code took around three years and was finally adopted in 1996, notwithstanding considerable resistance from many Bar associations and chambers. The Equality Code provides comprehensive and detailed advice to chambers of good practice on all equal opportunity issues. It covers all areas of practice and is being monitored for compliance by means of checks by the Bar Council.[57]

The adoption of the Equality Code represents a significant achievement for the Bar. It is a recognition of the fact that there are institutional barriers to the progression of women and other discriminated groups, what the Code refers to as 'inherited practices and assumptions'. Accordingly, the Code recognises that many existing practices at the Bar may constitute indirect discrimination, and that it is not just up to individual women to seek to change things and to seek to advance themselves, but that specific action needs to be taken at an institutional level to ensure real equality of opportunity. The adoption of the Equality Code is, of course, only a first step. The Code must now be implemented and complied with throughout the profession. There is much work on this to be done as by mid-1998, it appears that only around 50 sets of chambers had taken steps to implement the Code.[58] An attempt by the Bar Council to assess the extent to which the Code had been implemented was rendered largely ineffective due to the very low response rate, leading the Bar Council to contemplate a policy of 'naming and shaming' those chambers which have not responded.[59] Thus, strict monitoring by chambers, individuals and the Bar's management bodies is required, as minimal, or no, adherence will not be sufficient. Only if the Code is able to engender cultural change at the Bar, and develop an ethos of equality, will it be judged a success.

CONCLUSIONS

This chapter has developed the analysis from chapter five on access to the Bar and has considered many of the disadvantages which still face women barristers. It was also noted that the position is improving, with the implementation of the Equality Code representing a formidable task for the Bar. Heather Hallet QC hopes that her election as the first woman to chair the Bar Council will 'send out the message that the Bar is open to all – and that it is possible to be a working mother and get to the top'.[60] Only once the Bar has tackled the systemic and institutionalised discrimination present at the Bar will this become true for all but a few of the most talented individuals. Change is a slow process and constant vigilance and pressure is required to achieve lasting improvement. And change there must be. Barristers play a crucial role in the implementation and development of the law, and women must and should play a central part of this process. Furthermore, it remains the case that the judiciary is still drawn largely from the ranks of the Bar. Thus, in order to ensure a more representative and accountable judiciary, women must have equal access to, and success at, the Bar.

PERSONAL TESTIMONIES OF WOMEN BARRISTERS

Barbara Hewson, Littman Chambers, 12 Gray's Inn Square

Barbara Hewson was a founder member of the Association of Women Barristers, its Joint Vice-President in 1996–98, and has been at the forefront of the changes which have been taking place at the Bar. Her practice, in a wide field of European, employment and public law, involves many high profile cases, most recently those relating to court-ordered caesareans. She describes below her route to the Bar and her experience of working towards change ...

'I was called to the Bar in 1985, at a time when the number of women being admitted to the profession increased markedly. My background was conventional: a Catholic girls' boarding school in Sussex, followed by Trinity Hall, Cambridge (famous for producing one out of every ten judges, it was said). My interest in the Bar had been sparked at an early age from reading Patrick Hastings' marvellous *Cases in Court* and Edward Majoribanks' *Life of Marshall Hall.* I had also a passion for the works of Oscar Wilde: his fate at the hands of the English legal system gave me an insight into the sometimes terrible workings of British justice. However, I had taken the gamble of eschewing law at Trinity Hall and reading English; with the result that I then had to do a 'crash' course in law in London.

The thought of a daily diet of horror stories, emotional and physical, frankly turned me off crime or family law. So I went into the Chancery Bar, finding chambers in Lincoln's Inn in 1987. I enjoyed Chancery practice. There was more respect for your brain power than concern for what gender you were. I was never asked questions at interviews about my marital status or family plans. What really counted was one's ability to analyse a knotty legal problem and to present one's client's case effectively in court – either before a stern-faced Chancery judge with a keen eye for procedural improprieties, who did not suffer fools; or before case-hardened county court judges who were occasionally notorious for their temper. I only came across a woman judge once.

Having been acquainted with both theoretical and practical feminism at university, in due course, those feminist instincts reasserted themselves, and with the confidence that one acquires after having established oneself in practice, I became involved in the setting up of the Association for Women Barristers in 1991. This was considered rather daring at the time; although, as we kept pointing out, women solicitors had had their own Association since 1919. We had a clear agenda: to get the Bar Council to improve opportunities at the Bar for women.

I later moved chambers – to an EC/Public law set in Gray's Inn – and was elected to the Bar Council on an equal opportunities ticket. I found the experience of institutional politics fascinating. Stephen Sedley QC (now Mr Justice Sedley) chaired the new Sex Discrimination Committee, and we were immediately busy: developing a maternity leave policy, considering complaints of discrimination brought to the attention of the Bar's new Equal Opportunities Officers, considering the findings of the research we had commissioned, the *Without Prejudice?* research. The research gave hard evidence of disadvantage faced by women, and made recommendations for change. Many of these recommendations were incorporated into the Bar's new Equality Code which met with considerable resistance from some other Bar Associations.

Judicial appointments and silk remain a real problem area. One difficulty is the insistence that in order to be a 'proper' judge, you need to be in private practice. Another is the use of the pool of QCs as the filter for selection barristers as High Court judges. Since women form only a tiny percentage of this pool (about six per cent), the number of women on the High Court Bench – and above – is set to remain unnaturally low. Finally, there is the system of 'secret soundings'. Given that most judges are men, there is a tendency to select in their own image, based on loose personal impressions, rather than rigorous assessments based on objective criteria.

Institutional politics won't let you go; you have to walk away from it. In the end I decided that, riveting though it was, I wanted to spend more time on my practice. Although I think opportunities for women are as good now as they have ever been, I think that the future for the Bar in private practice is perhaps less inviting. Solicitors are increasingly doing the work that used to be the staple of junior barristers: settling pleadings and affidavits; interlocutory court appearances. More and more, they prefer to advise in-house rather than coming to counsel. Legal Aid is diminishing. The traditional chambers model – a loose conglomerate of sole traders who share expenses but not profits, who cannot form partnerships or employ each other, and who are 'managed' by clerks – is becoming unworkable in the late twentieth century. I think that the environment which women barristers coming into the profession will encounter will be increasingly more like a firm of solicitors.

So far as feminist legal practice goes, my work now includes equality issues: from sexual harassment and maternity rights at work, to the rights to bodily integrity of pregnant women in recent 'forced Caesarean' cases. Issues of fundamental human rights can arise in a wide variety of contexts. I have also found the intellectual rigours of my early Chancery experience to be an invaluable training for the increasingly complex issues which now face me in my practice.

The use of EC law to promote women's equality in the workplace, and the forthcoming incorporation of the European Convention on Human Rights, have transformed and will continue to transform the way in which issues affecting women are litigated, and the way in which women are perceived by the courts. It is vital that women coming to the Bar should have an awareness of human rights law, of EC law and of contemporary feminist legal scholarship: ultimately, it is only by formulating the right questions for the court, and arguing them in the right way, that legal questions of the utmost importance for women can be adjudicated upon in ways that will advance, not diminish, women's claims to equality and justice.'

Najma Khanzada, 1 Pump Court

Najma is a junior barrister practising almost exclusively in criminal defence work and has done so since she was taken on as a tenant at 1 Pump Court in 1994. She describes below her experiences and perspectives on gender and racial discrimination at the Bar, and how she has learnt to deal with such difficulties and continue to enjoy her work ...

'Like many, my route to pupillage and tenancy was not a direct one. I gained along the way a philosophy and political theory degree, various jobs in the voluntary sector relating to women and domestic violence, a law degree, part-time law lecturing posts on the external London LLB course, and para-legal and outdoor clerking work for solicitors' firms. Although indirect, this route was marked by a single-minded and focused determination that I wanted to be a barrister.

Although I had a strong desire to be a barrister, my reasons for that choice were less clear. If I am honest, it was only when I was forced to articulate them at interviews that I gained some insight into my own decision. I wanted independence; I wanted a profession that reflected my social welfarist politics; I wanted to be an advocate. At a superficial level, I wanted to stride around in sharp suits and be paid for being aggressive.

The temptation is, when asked about one's experiences of entry to the profession and establishing a practice, to pretend that the process has been a smooth one. The Bar forces one to take on a veneer of self confidence and resilience that is often a lie. To admit that there are or have been problems is to admit that one has a problem. The reality is that the work is problematic and stressful. From my experience as a criminal barrister, there are problems and stresses which uniquely affect women. What has become clear to me in my gravitation towards the Bar, and in practice, is that barristers do suffer discrimination on the basis of gender and race.

The problem is the subconscious attitudes and preconceptions of clients, be they lay or professional, and of the judiciary. It arises from the way women are perceived and the stereotypical picture of what constitutes a good criminal practitioner. This is caused and exacerbated by the nature of criminal defence work itself. Criminal defence work provides a subtle blend of grime and glamour. My great grandfather ran a wrestling school, and I was once told that my work was no different from his. The grime comes from figuratively wrestling in the mud with the diffident, abusive, mentally ill or physically threatening client; the glamour in fighting as a pugilist in front of a captive audience of the jury with bravado and nonchalance. Most clients are men and to date most criminal barristers are men. Clients want 'bruisers', they want big burly wrestlers with presence.

The ways this attitude impinges on one's working life are diverse. Solicitors are less inclined to brief junior women practitioners for fear of disappointing the client. This decision may well not be conscious, but as a woman you are burdened with the task of convincing a solicitor that your are competent. Moreover, the net result is that a male colleague is more likely to be briefed instead.

As a junior practitioner, your first contact with a client is often through a cell wicket in a magistrates' court. Those first moments of meeting stamp their mark on your relationship with that client. The response from the other side of the cell door is often to revert to 'chatting up' the representative. Sometimes it is abusive; I have been assaulted once and threatened with assault on a few occasions. It is often critical. There are of courses simple techniques to forestall the negative response, but the first time it happens it is a shock.

Again as a junior practitioner one is often a sitting duck for the jaded or bullying tribunal. To take a risk in advocacy, to divert from the text book style of cross examination or closing speech, can make one vulnerable to badgering and unnecessary intervention from the bench. This has rarely anything to do with one's competence and a lot to do with an individual judge's irritations and difficulty in treating a female or non-white advocate fairly. Styles of advocacy are changing and black working class women should not be forced to talk as Rumpole did.

The knock-on effect of the badgering judge is that client relationships are undermined. They do not see their 'bruiser' fighting another battle. They see their woman barrister being humiliated. The ultimate effect of this problem is that one internalises the doubt that others have of you. The discrimination ceases to be an external problem and becomes your problem. You are the problem.

On a practical level there appears to be little that can be done about the attitudes around us. It therefore seems important to be able to share and thus unburden oneself of the negative effects of working in an often hostile environment. The first step has to be to form informal or formal support network groups in specific areas of practice. To this end a group for women criminal advocates has been established, Women in Crime. It provides a forum for women solicitors and barristers to share common experiences and to meet socially. Further, on a practical level, it aims to organise training in both advocacy and presentation skills.

A further practical step, although sometimes impossible, is to gain at least verbal acknowledgement by your clerks and colleagues that women have specific difficulties in establishing a practice in crime. My personal experience of this has been perhaps unusual and fortunate. My chambers is notable for having a large proportion of women members. It has strived to adopt and implement equal opportunities policies and recently won an award from *The Times* for this work. As a working environment it is supportive of its individual members and of the values which underpin a predominately legal aid and social welfare practice. Problems of women in crime are regularly addressed in meetings and allocation of work is to be monitored through the ACE fee system in order to identify trends in briefing decisions. I may never be the wrestler that my great grandfather was, but I, like many, deserve the chance to try.'

Geraldine Andrews, Essex Court Chambers

Geraldine practises in the area of commercial law which she finds consistently stimulating. She describes below her career path, and offers her thoughts on the position of women at the Bar ...

'Until I was about 14, I envisaged no career other than that of a professional musician. However, when it became clear that I would have to jettison most of my studies in order to pursue music full-time, I began to have second thoughts. At about this time, (the mid 1970s) a remarkable woman came to my school to speak about legal careers. She was a solicitor with a local firm, and she spoke about the law with such enthusiasm and commitment that I became intrigued. I explained to her that I had no family connections in the law and moreover, had heard that the Bar in particular was a very male-dominated profession. However, the combination of the performance element and the academic challenge attracted me. She was extremely encouraging, and told me where I could obtain more information. She also said that I had nothing to lose by aiming high, provided that I was also realistic – advice I often now find myself passing on to aspiring lawyers.

Before long, I knew that I had found my true vocation. After spending three happy years at King's College London, I graduated with First Class Honours and set about looking for a pupillage. By this time, I knew that I wanted to pursue a career in commercial law, and so I applied to the set of Chambers recommended to me as the top in that field, which was then located at 4 Essex Court. I was fortunate enough to be offered a pupillage after an informal interview – the selection process has since become far more sophisticated! In those days, there were no Chambers' pupillage awards and it was not easy to finance this part of my education, particularly as my father had just died. I subsisted on a mixture of grants and a scholarship from Gray's Inn and commuted from home to save money.

After Bar exams, I returned to King's to take an LLM and then, on completion of pupillage, I was taken on as one of three new tenants. I have practised from these Chambers (now known as Essex Court Chambers) ever since. In that time, with the aid of a first-rate clerking team, I have built up a successful practice and I believe that I have gained a good reputation in my chosen field. I have also been extremely happy here. At first, it was somewhat disconcerting to be the only woman tenant practising from Chambers full-time (we had a female door tenant who has since gone on to become a judge). However, it was apparent to me from the outset that this was not due to misogyny, but rather to the absence of interest in the commercial field being shown by suitably-qualified women, and a resulting lack of suitable applicants. Over the years, this has changed, and there are now eight women tenants in Chambers.

I am aware that I have been very fortunate, and that I took a number of risks which paid off, but that is probably the only way in which I could have succeeded in my aims. I have never personally experienced discrimination at the Commercial Bar. When one practises in a specialist area, discrimination and harassment are usually neither expected nor encountered. Indeed, one of the best aspects of the profession is the pleasant working environment in Chambers and the cordial relationship with my colleagues, many of whom have become very good friends. I am aware that others whose practices are more general in nature have been less fortunate, although times have changed considerably, and I would never deter anyone from coming to the Bar on the basis that they will be picked on by men, or made to do family or criminal work by their clerks. Nowadays I do not regard a female applicant for pupillage or a tenancy as being under any disadvantage compared with her male competitors.

What I *would* say to any woman contemplating a career at the Bar is that it requires enormous self-confidence, resilience and stamina, and that she should not even take the first step unless she is absolutely committed to it. You probably still need to be as good as, and preferably better than, your male colleagues in order to get noticed, at least at the beginning. Yet it is a big mistake to try and turn yourself into a 'surrogate man'; your femininity is a tremendous asset which can, and should, be used to your advantage wherever possible. It is imperative to be true to yourself.

The demands of the work wreak havoc with your social life; the hours are long, and you regularly work at evenings and weekends and sacrifice holidays, especially in the early years. Some women have managed to fit in the demands of a family, and they have my complete admiration. I have remained single, and regard it as part of the price which I have had to pay for climbing the ladder towards the top of my profession.

You may also struggle for years without making much money; sometimes you will find it hard to subsist. No-one should come to the Bar simply because it appears to be lucrative. You will find every penny is hard earned, even if you are fortunate enough to be a high-flyer. However, the rewards for your dedication are great: there is nothing quite like the feeling of personal fulfilment when you sit down after making a successful speech or concluding a devastating cross-examination.'

NOTES

1 Quoted in Richard Abel, *The Legal Profession in England and Wales*, Blackwell: Oxford, 1988, p 85.
2 TMS Consultants, *Without Prejudice? Sex Equality at the Bar and in the Judiciary*, Bar Council and LCD, 1992, pp ii–iii.
3 Paul McCann, 'Rumpole sentenced to an early retirement', *The Independent,* 13 January 1998.
4 All statistics in this chapter were given to the author by the Bar Council, unless otherwise noted.
5 Supra note 2, p 8.
6 Frances Gibb, 'Sexual bias is still a problem, says a Bar pupil', *The Times*, 27 April 1998.
7 Quoted in Clare Dyer, 'Women QCs "enjoy reverse bias"', *The Guardian*, 30 September 1996.
8 LCD Press Releases, 'Queen's Counsel 1996', 94/96, 4 April 1996; 'Queen's Counsel 1997', 61/97, 26 March 1997; 'Queen's Counsel 1998', 86/98, 9 April 1998.
9 Barbara Hewson, 'You've a long way to go, baby ...', 146 (1996) *New Law Journal* 565–566, p 565.
10 Ibid.
11 Figures quoted by Lord Chancellor, Speech to 1998 Woman Lawyer Conference, reproduced at: http://www.open.gov.uk/lcd/speeches/1998/1998fr.htm.
12 Supra note 8. 1992 was the first year that statistics on ethnic minority background were maintained.
13 Figures quoted by Lord Chancellor, Speech to 1997 Minority Lawyers Conference, reproduced at: http://www.open.gov.uk/lcd/speeches/1997/speechfr.htm.
14 Supra note 9.
15 For further information on the AWB, see appendices five and six.
16 The Royal Commission on Legal Services, Cmnd 7648, 1979, para 35.29, p 501.
17 Frances Gibb, 'Working mum reaches the top', *The Times*, 6 January 1998.
18 Quoted in Anthony Doran, 'Lawyers hold no brief for equality code', *The Daily Mail,* 7 November 1995.
19 See chapter four under the heading 'Explaining it away'.
20 Hazel Williamson QC, 'Networking and gaining promotion', Handout, 1995 Woman Lawyer Conference.

21   Helena Kennedy, 'Women at the Bar', in Robert Hazell (ed), *The Bar on Trial*, Quartet Books: London, 1978, p 160.

22   *The Independent*, 26 November 1990.

23   The situation of employed barristers will be more akin to solicitors in that they will be entitled to statutory maternity rights and any additional provisions offered by their employer.

24   Young Women Lawyers (YWL) Press Release, 'Women Law Students Anticipate Discrimination', 3 February 1998, and quoted in YWL Newsletter, January 1998. For further information about the group Young Women Lawyers, see appendices five and six.

25   Supra note 2, p 11.

26   Bar Council, *Annual Report 1997*, p 17.

27   Lecture to the AWB, reproduced at http: //www.open.gov.uk/lcd/judicial/women.htm.

28   S Shuaib, 'Discrimination still exists', *The Independent*, 21 June 1991.

29   John Flood, *Barristers' Clerks: the law's middlemen*, Manchester University Press: Manchester, 1993.

30   Reported in Yasmin Alibhai-Brown, 'Breaking the colour bar', *The Guardian*, 7 April 1998.

31   Supra note 2, p 10.

32   See Laura Cox and Barbara Hewson, 'Equality of Work Distribution in Chambers', *Counsel*, September/October 1996.

33   Joanna Shapland and Angela Sorsby, *Starting Practice: work and training at the junior Bar*, Institute for the Study of the Legal Profession: University of Sheffield, 1995.

34   The report is internal and was not published, but details are considered in Barbara Hewson, 'Sexual Harassment at the Bar – A recent problem?', 145 (1995) *New Law Journal* 626–627.

35   Quoted in Clare Dyer, 'Women bar students "are regularly offered traineeships for sex"', *The Guardian*, 4 May 1995.

36   Supra note 34.

37   She can activate a Chambers' grievance procedure, raise the matter with, and possibly make a complaint to, the student officer of an Inn, or seek advice from one of the Bar's Equal Opportunities Officers. A complaint can be made to the Bar's Professional Conduct Committee (PCC), an action can be brought in the county court or industrial tribunal, and where relevant a complaint may be made to the police.

38   Clare Dyer, 'Sex pest judge quits the bench', *The Guardian*, 3 March 1998.

39   Laura Cox and Barbara Hewson, 'Dealing with Harassment – A Practical Guide', *Counsel*, May/June 1996.

40   Sleeve cover, Nicola Williams, *Without Prejudice*, Headline Feature: London, 1997.

41   Anesta Weeks, Conference Handout, 'Black women at the Bar', 1995 Woman Lawyer Conference.

42   For more information about the SBL, see appendices five and six.

43   Supra note 2, p 20.

44   See chapter seven under the heading 'Partially open: circuit judge, district judge and assistant recorders'

45 The criteria are detailed on the LCD's web site: http://www.open.gov.uk/lcd/judicial/silk3.htm.
46 Quoted in supra note 17.
47 Supra note 9, p 565.
48 Ibid.
49 Editorial, 'Less tinkering, more action', *The Lawyer*, 1 April 1997.
50 Supra note 9. In addition, it may well reduce the costs of the legal system: see, Peter Reeves, *Silk Cut: Are QCs really necessary?*, Adam Smith Institute: London, 1998.
51 See chapter five under the heading 'Joining Inn'.
52 Reported in *Counsel*, July/August 1997, p 5.
53 Ibid.
54 Supra note 16, para 35.30, p 501.
55 Supra note 9, p 565
56 Quoted in Colleen Graffy, 'Coming to terms with keeping terms', *Counsel,* March/April 1997, p 11.
57 Frances Gibb, 'Bar to launch spot checks on compliance with equality code', *The Times*, 12 April 1997.
58 Supra note 6.
59 *The Lawyer*, 5 June 1998. Only 92 of 420 sets responded.
60 Supra note 17.

Chapter 7
Judging women

'It is unlikely that the judicial appointment system offers equal access to women or fair access to promotion to women judges ... The system depends on patronage, being noticed and being known...'

Without Prejudice? Sex Equality at the Bar and in the Judiciary, 1992[1]

'The consequences of this lack of women in the senior judiciary reverberate through every level of society, reinforcing the de facto status of women as less authoritative then men, and the perception of their experience and contribution as less valuable.'

Chair, Association of Women Barristers, 1996[2]

It is a sad irony that the mythical symbol of justice, the blind-folded figure holding the scales of justice, which appears above our courts, is female, yet women as judges, the dispensers of justice, are few and far between. The first woman was appointed to the US Supreme Court in 1981 and to the Supreme Court of Canada in 1982: but there is no woman, nor has there ever been one, in the House of Lords, the UK's highest court. The most senior woman judge in the UK is Lady Justice Butler-Sloss, who sits in the Court of Appeal. The judge is the quintessential authoritative figure of the law, and, as we have considered in previous chapters, women are not generally viewed as authoritative, whether we talk of women law professors, women barristers or women judges. In the words of the Australian law professor Margaret Thornton, women remain 'fringe-dwellers of the jurisprudential community'[3]: this is nowhere more evident than in the judiciary.

The gender composition of the judiciary in England and Wales is the subject of the first part of this chapter, which then goes on to examine the main features of the judicial appointments process. In the light of these discussions, some of the proposals to reform the appointments process are considered. The final part looks at some of the reasons why we are so interested in appointing women judges, focusing on the question of whether more women judges will make a difference to the law and legal profession. The chapter concludes with the personal testimonies of four women judges which expand on and exemplify some of the themes identified below.

JUDGING THE JUDICIARY

We can see from table 7.1 that there are very few women in any of the judicial offices. The comparison with figures from 1989 shows that while there has been some very welcome progress over the last ten years, it has been limited. After Lady Justice Butler-Sloss, the most senior women judges are to be found in the High Court which hears the most serious civil and family law matters. High Court judges are drawn from a wide variety of legal backgrounds including the circuit bench, straight from practice or, more unusually, from academia. Mrs Justice Arden is not only the sole woman judge in the Chancery Division of the High Court, but she was the first woman ever to be appointed to that division. Mrs Justices Ebsworth, Smith and Steel are judges of the Queen's Bench Division, with Mrs Justices Bracewell, Hale and Hogg in the Family Division, making a total of seven women High Court judges.

The number of women circuit judges has almost doubled in the years since 1989. Although this is a significant increase, the percentage of women remains much the same, as the overall number of circuit judges has increased, with numbers of women increasing proportionately. Circuit judges normally sit in both civil and criminal cases, with some judges being specifically authorised to hear family law matters. Circuit judges will hear many of the most serious criminal charges, including murder and serious sexual offences, as well as the majority of fraud cases. In civil matters, circuit judges will generally hear those cases of a value less than £50,000. The Lord Chancellor will normally only consider applicants for the office of circuit judge who are aged between 45 and 60, who have been recorders for at least two years, and there is an expectation that QCs applying will have taken silk more than three years ago.

The office of assistant recorder is often seen as the first rung on the judicial ladder, and it is at this level that the greatest advances in the representation of women have been made. Assistant recorders usually sit in the Crown Court hearing criminal matters, although some hear only civil and family matters. Applicants for assistant recorderships must usually be aged between 35 and 53, and from this office it is possible to become a recorder, then perhaps a circuit judge or (deputy) High Court judge. In 1989 there were only 38 women recorders and 25 assistant recorders, figures which had increased by 1998 to 69 and 61 respectively, with women now representing 17 per cent of assistant recorders. In the last competition, 16 per cent of applicants and 22 per cent of successful applicants were women.[4] (No such figures are published regarding circuit judges.)

Table 7.1: Number of women judges in 1998 and 1989[5]

	Women bar. sol.		Total women	Men bar. sol.		Total men	% women 1998	% women 1999
Lords of Appeal in Ordinary (House of Lords)	-	-	-	12	-	12	0%	0%
Lords Justices of Appeal (Court of Appeal)	1	-	1	34	-	34	3%	4%(1 woman)
High Court Judges	7	-	7	89	1	90	7%	1% (1 woman)
Circuit Judges	25	5	30	446	71	517	5%	4% (17 women)
Recorders	65	4	69	722	71	793	8%	5% (38 women)
Assistant Recorders	43	18	61	240	53	293	17%	5% (25 women)
Ass. Recorders in training	12	3	15	66	7	73	17%	-
District Judges	1	37	38	3	296	299	11%	4% (7 women)
Deputy District Judges	5	69	74	11	612	623	11%	-
Stipendiary Magistrates	12	2	14	26	51	77	15%	13%(8 women)
Total			308			2811	10%	

The percentage of women district judges has almost trebled between 1989 and 1998. The vast majority of district judges are solicitors, largely due to the civil nature of the work and the fact that solicitors tend to have less experience of court work. District judges dispose of over 80 per cent of contested civil litigation in England and Wales, and are usually aged between 40 and 60. Unfortunately, however, the ranks of district judges are not seen as natural breeding grounds for more senior judicial appointments for which some form of criminal experience is largely required. Nonetheless it is possible to progress from the office of district judge to assistant recorder and onwards. No figures are published relating to the proportion of women applicants and appointments, although this information should be readily available.

Although the number of women at recorder and district judge level has increased, there has been little improvement regarding stipendiary (legally qualified) magistrates. There were eight women stipendiary magistrates in 1989 (13 per cent), and only 14 (15 per cent) in 1998. In contrast, figures for 1994 show that 46 per cent of lay magistrates were women, perhaps reflecting the appointments policy that the magistracy should be representative of the wider community.[6]

It is a commonly held view that the number of women in senior positions will improve given time, and this is no less so with the judiciary. Indeed, the former Lord Chancellor, Lord Mackay, argued that: 'As more women progress through the profession, it is to be expected that the numbers of women within the judiciary will increase.'[7] Such reasoning is, however, far too simplistic. The *Without Prejudice?* research revealed that women are disadvantaged at the Bar and in the judiciary in a number of ways which means that they are not succeeding as might have been expected. A Law Society study in 1991 similarly found that women had a lesser chance of becoming a judge than men.[8] Kamlesh Bahl, chairwoman of the Equal Opportunities Commission, has rebutted the Lord Chancellor's assumption by pointing out that women have been entering the profession in significant numbers for many years. She has referred, in particular, to the teaching profession where, despite a high volume of women, women remained under-represented in the upper echelons, and continued: 'the law is no different from other professions. We cannot count on mere 'mass' to ensure change at higher levels. What we need is concerted action to dismantle barriers to judicial appointments of women'.[9]

Table 7.2: Number of ethnic minority judges in 1997 and 1994[10]

	No. of ethnic minority judges	Total	1994 (number)
Lords of Appeal in Ordinary	-	0%	-
Lords Justices of Appeal	-	0%	-
High Court Judges	-	0%	-
Circuit Judges	6	558(1.1%)	4
Recorders	14	929(1.5%)	11
Assistant Recorders	9	323(2.8%)	8
District Judges	3	335(0.9%)	1
Deputy District Judges	13	697(1.9%)	7
Total	45	2842(1.6%)	31

Action is also required regarding the representation of ethnic minorities on the Bench. Table 7.2 shows that the most senior ethnic minority judges are circuit judges, of which there are only six, and that, as with the representation of women, the greatest number of ethnic minority judges can be found within the ranks of assistant recorder and deputy district judge. What is not known is how many of these ethnic minority judges are women, as published figures are not broken down in this way. Table 7.2 also shows that improvements have been made between 1994 and 1997. In particular, four per cent of successful candidates for the positions of assistant recorder and circuit judge in 1996–97 are from ethnic minorities.[11]

APPOINTING JUDGES

There are a vast array of different judicial offices, with around 600 appointments being made each year.[12] These range from the most senior judicial offices in the House of Lords or Court of Appeal, to the extensive number of tribunal appointments. The appointments process for such a large number of offices will necessarily be complex. However, valid criticisms can be made of the procedures for appointing judges on the basis that they are shrouded in mystique, are not fully open and transparent, rely too heavily on the views of the present judiciary and senior members of the profession, and are masterminded by a government minister, the Lord Chancellor. Debate regarding why there are so few women judges has tended to focus on the judicial appointments process, resulting in many reforms of the procedures since the early 1990s. Debate has moved on and discussion now centres on the extent of reforms. The first section below will examine the appointing methods for the principal judicial offices, for which broadly three different procedures are in operation. I have described these processes as: partially open, partially closed and closed.

Partially open: circuit judge, district judge and assistant recorders

Appointment to the offices of circuit judge, district judge and assistant recorder is by application only, with vacant positions being advertised. The first stage of the process involves consultations with senior members of the profession, otherwise known as the process of 'secret soundings' (see further below). On the basis of the information in the application form and the recommendations of consultees, selection for interview is made, and interviews are held by members of the Judicial Appointments Division of the Lord Chancellor's Department (LCD), a lay person (drawn from the Lord Chancellor's Advisory Committees) and a serving judge of appropriate seniority. However, it is not the interview panel which appoints candidates. The panel simply places its views before the Lord Chancellor, together with the comments arising from the consultations with the judiciary and senior members of the profession, who then decides who to appoint (assistant recorders and district judges), or who to recommend for appointment to the Queen (circuit judges and recorders).

Hence, the process is a mixture of a more usual open and transparent process of application and interview, with closing dates for applications and published timescales within which appointment decisions are to be made, and the process of soundings and consultations which are not open and transparent and in respect of which the applicant has little opportunity to respond. And finally, the appointment ultimately rests with a member of the government, the Lord Chancellor.

Partially closed: the High Court

Although the process described above may be open to criticism, it is far more progressive than that in place for the more senior judicial appointments. The position of High Court Judge was advertised for the first time in 1998. However, unlike the procedures for those offices considered above, appointment to the High Court does not *require* an application. The Lord Chancellor has reserved the right to offer appointments to those who have not applied, on the basis that he 'must have a way of reaching those who are able, but are unwilling, to press their own claims'.[13] If an individual is willing to serve as a High Court judge, why should they not be required to apply in the usual manner? Do we really want such shrinking violets in the High Court? It may well be the case that some individuals require persuasion, but this should be persuasion *to apply*.

The Lord Chancellor states that an openly advertised selection procedure ensures 'a fair and equal chance' for applicants.[14] But how can this be the case if the appointments process is not the same for all? There will always be seen to be a hierarchy of those who did not need to apply (and are therefore obviously good), and those who did apply (those seen as not so good as they did not think they would be invited in any event). In addition, before an individual is invited to become a High Court judge, the consultation process will have been completed, and what then if the invitee refuses the offer? What a waste of time and effort in the collection of the consultation responses. Finally, it will be more difficult to monitor the appointments process. It would certainly be very interesting to see statistics in the years to come regarding those women and men who were offered a position without applying, compared to the proportions of women and men applying and being appointed.

Furthermore, the criteria for appointment to the High Court state that prior sittings in a judicial capacity, either as a recorder or deputy High Court judge are 'highly desirable', but not necessary. Although appointment to assistant recorder is made on the basis of the relatively open procedure considered above, there is no appointments process for deputy High Court judges; it is by invitation only.[15] The Lord Chancellor, in making his recommendation to the Queen, takes soundings from senior members of the profession and judiciary, and in some cases will have the information contained on an application form. But there are no interviews, and no lay element in the selection process.

There is no reason why the introduction of the more transparent appointments mechanisms for other judicial positions could not be extended to the High Court bench. The Lord Chancellor, in announcing the first advertisement for the High Court, said that he was 'particularly keen that eligible women and ethnic minority lawyers should give this opportunity serious thought'.[16] However, he does not appear to consider that the moves to improve the appointments process lower down the judicial ladder apply to the senior judiciary, even though it is arguable that they hold more important positions and therefore scrutiny of their appointments should be all the greater.

Closed: Lords/Ladies of Appeal in Ordinary and Lord/Lady Justices of Appeal

The most senior judicial appointments, to the House of Lords and Court of Appeal, remain closed. There are no job descriptions, no selection criteria, no advertisements and no lay participation. The LCD simply states that Lords of Appeal in Ordinary are 'lawyers of great distinction who have proved themselves capable of handling the most important and complex issues which arise in appeals'. And, appointment to the Court of Appeal is 'from the ranks of the most experienced High Court judges'.[17]

What is more, the appointment is made by the Queen, on the recommendation of the Prime Minister, who takes advice from the Lord Chancellor. Before tendering advice, the Lord Chancellor will consult senior members of the judiciary. The last two Lord Chancellors, Lords Mackay and Hailsham, are both on record as stating that their respective Prime Ministers (John Major and Margaret Thatcher) did not always take their first recommendation.[18] Thus, appointment to the two most senior courts is a *political* appointment, decided among the Prime Minister, Lord Chancellor and the forty or so senior judges. This is a lamentable process for the appointment of the most senior judges who are supposedly independent of the executive and has been described as 'nothing less than naked political control over appointments to the most senior levels of the judiciary'.[19]

Although the number of potential candidates is necessarily small, this does not reduce the requirement for openness and accountability. The Equal Opportunities Commission (EOC) and the Association of Women Barristers (AWB) have both recommended that the appointment and selection procedures now applied in respect of the less senior judicial offices be applied to all judicial positions.[20] Parliament's Home Affairs Committee has recommended that job descriptions and selection criteria be formulated for all judicial positions 'without delay', and the Committee had 'some qualms' about the role of the Prime Minister in the appointments process and questioned whether this practice should continue.[21]

'Secret soundings'

The fundamental basis of each of the appointing procedures described above is the consultations which take place with senior members of the legal profession and the judiciary, which are known as 'secret soundings'. The Lord Chancellor has stated that he does not like the use of 'secret soundings' as it is, apparently, a 'sinister' term.[22] Its use, however, ensures that a sceptical mind is kept over the process. If it is the process of soundings that is one of the principal reasons why women are not as successful in taking silk or gaining judicial office as men, as is often argued, then the process may indeed be sinister.

The *Without Prejudice?* research in 1992 concluded that: 'It is unlikely that the judicial appointment system offers equal access to women or fair access to promotion to women judges ... The system depends on patronage, being noticed and being known.'[23] That this is the case is accepted by many judges themselves. Lord Justice Otton said at the Woman Lawyer conference in 1996 that so long as we have 'male consultors consulting male consultees, women are bound to be at a disadvantage'.[24] And, in 1992, Lord Bridge, a retired member of the House of Lords, agreed that when selecting a new judge, they looked for 'chaps like themselves'.[25]

The AWB sought to test this argument, in part, by hypothesising that the judges consulted tend to recommend for appointment to the High Court bench, barristers from their own chambers.[26] The AWB gathered information on all High Court appointments between 1986 and 1996 and on the judges of the High Court, Court of Appeal and House of Lords of the relevant consultation period. Only 58 of 227 sets of chambers were represented by the judicial appointees during the period of the study. Of those 58, seven sets produced no less than 30 judicial appointments. In other words, 29 per cent of all judicial appointees came from three per cent of sets of chambers in London, and two per cent of chambers in England and Wales.[27] Although the figures involved in this study are small, the results are indicative of the importance of 'being known'.

Where reliance is based on being known, this is inevitably more difficult for many women, especially in a profession dominated by a masculine culture and traditions. In 1996, the AWB recorded how some male barristers engaged in trials lunched each day, with the judge, in an all-male club. Women barristers were excluded except for certain occasions when they were allowed to enter as 'guests'.[28] This sort of exclusion of women is certainly not uncommon and ultimately creates not just a feeling of inequality, but the notion that women barristers are the exception and not the norm. Its effect is that they are less likely to come to the attention of consultees. The AWB has also suggested that Freemasonry, from which women are excluded, may have an adverse impact on women, as it is 'not unreasonable to suppose that there exists considerable opportunity for informal but important contacts and influence amongst men who practise Freemasonry'[29] and that this is 'damaging to public confidence in the impartiality of the judiciary and also to confidence in the impartiality of appointments to QC and the Bench'.[30]

The EOC also questions whether receiving comments from senior members of the judiciary which are not based on contemporaneous observation are indicative of a person's suitability.[31] Furthermore, given the predominance of men in the judiciary, on whose comments appointments are largely based, there is a risk that stereotypical assumptions will be made as to what constitutes 'female' as opposed to 'male' qualities and aptitudes, with potentially adverse consequences.[32]

These criticisms were raised in 1991 when solicitor Geoffrey Bindman wrote an article in the *Law Society Gazette* which suggested that the judicial appointments process, based on secret soundings, may contravene the Race Relations Act.[33] In response to the article, the Lord Chancellor sought counsel's opinion on whether this was a correct interpretation of the law, and on receiving a negative response, published a detailed rebuttal of the argument.[34] These questions about racial and sexual discrimination in the appointments process may be aired in the case brought in 1998 by barrister Josephine Hayes, who was chair of the AWB in 1997–98. Her action is against the Attorney General and suggests that the Government shows bias in favour of men when appointing lawyers to represent it in civil cases.[35] The complaint is that there is no method by which one can apply to, or be considered for, this work, and that selections are made on the basis of the same sort of secret soundings by which QCs and judges are selected. Applicants are therefore restricted to who is known to those with whom soundings are taken. Thus, this case raises significant questions about whether the secret soundings system complies with the present law. Since this action was brought, the Bar Council has began an investigation into the appointment of government counsel.[36] In addition, the Attorney General announced an inquiry into the method by which government lawyers are appointed, although he denied that there was a connection between this and the case brought by Josephine Hayes.[37]

Underpinning the judicial appointments process is that relating to QCs. We saw in the previous chapter that QC appointments are made solely on the basis of secret soundings: there is no lay element in the selection process and no interviews. Appointment to QC is a crucial, though not essential, step towards the bench. The AWB has argued that for as long as the numbers of women QCs remains low, the appointment to the High Court from the rank of QCs will disadvantage women.[38] The appointment procedures for QC have the effect, therefore, of acting as a filter. Accordingly, so long as appointment to QC is so criticised, appointments to the Bench must also be tainted. Making no findings of the method of appointment to QC, Parliament's Home Affairs Committee noted that if the appointment system for silk is flawed (as suggested in the previous chapter), then that 'route would not be wholly satisfactory for the appointment of judges'.[39]

Selection criteria

Although a first step towards a more open and transparent appointments process is the drafting of selection criteria against which a candidate is to be judged, this is only a first step. Difficulties begin with how to assess the candidate against the criteria, and the nature of the criteria in the first place. Accordingly, it is not just the fact of taking soundings that is so criticised, but also the criteria against which the applicants are supposedly judged. For each judicial office, excluding the Court of Appeal and House of Lords, selection criteria have been drawn up by the LCD which relate to legal knowledge and experience, skills and ability, and personal qualities.[40] The specifications for legal knowledge and experience vary considerably between the posts, as would be expected. In terms of skills and ability and personal qualities, however, the criteria are the same for each post. The requirements state that successful candidates will have intellectual and analytical ability, sound judgment, decisiveness, communication skills and authority. As for personal qualities, it will be necessary to demonstrate integrity, fairness, understanding of people and society, maturity and sound temperament, courtesy and humanity and commitment.

It is the policy of the LCD that those members of the judiciary and profession who are consulted over applicants for the judiciary, judge the candidates against the criteria specified above. However, exactly how one is to judge some of these criteria is a mystery. More importantly, however, is the danger of gendered assumptions and stereotypes forming the basis for the assessment of the individuals. Solicitor Kamlesh Bahl, who is chair of the Equal Opportunities Commission and Deputy Vice-President of the Law Society 1998, has criticised the selection criteria of 'decisiveness' and 'authority' as being subjective and leading to judgments based on individuals' preconceptions.[41] The great danger in an area such as the judiciary, she suggested, is that it has always been largely seen as a male area of work, so perceptions of what makes a good judge – and what is 'authority' and 'decisiveness' – are also likely to be male.[42]

We noted at the beginning of this chapter that it is precisely the sort of criteria as 'authority' that women are most often seen as lacking. Authority is an elusive concept, dependent more on what others think, than on one's own qualities. In other words, if there is a court room full of individuals who consider that women should not be on the Bench but at home, the woman judge will not have much 'authority'. But, this has little to do with the qualities of the individual woman judge. In fact, to rely on the views of others in this way is to condone age-old prejudices. Thus, even where soundings are taken in relation to the specified criteria for the judicial appointment, problems are likely to arise. We also saw this phenomenon in terms of the criteria of 'standing' in relation to silk, and the use of 'commitment' as a criteria for professional success was also doubted in that it can have a particularly gendered interpretation.[43]

FROM RHETORIC TO REALITY

Refining the system

There have been many calls for reform of the judicial appointments process in recent years. The following section will examine potential reforms aimed at refining the present system to make it more egalitarian, with more large-scale proposals being considered thereafter. The first step, as outlined above, must be the introduction of the more open selection procedures adopted for circuit/district judges and assistant recorders for all judicial appointments, no matter how senior. Accordingly, therefore, advertisements must be placed, selection criteria drafted and job descriptions promulgated. Lay participation in the process, together with use of meaningful interviews, must be introduced. It may even be that when appointment to the most senior courts, such as the Court of Appeal and House of Lords, is announced, and applications invited, public debate takes place over the various candidates, on their track records, declared opinions and background.[44]

Part of any selection process must be information on the applicants' performance. The EOC has suggested that a structured monitoring program should be introduced for those on the lower rungs of the judicial ladder. The aim is for assessment to be more objective and less reliant on the reminiscences of judges.[45] It is encouraging that the Lord Chancellor has announced the introduction of a form of appraisal for newly appointed assistant recorders and deputy district judges, although it remains to be seen what format this will take and whether it will comply with standard equal opportunities requirements.[46]

Even if maintaining the system of consultations, the least that should be done is for those being consulted to be informed of the available research and literature on the difficulties inherent in making such assessments. Accordingly, the EOC has recommended, at a minimum, that equal opportunities training be taken by those engaged in the consultation process so that they may be made more aware of their possible stereotypical attitudes.[47] Similarly, the AWB has criticised the fact that the senior judiciary were not briefed on the findings of the *Without Prejudice?* research which raised so many concerns regarding judicial and QC appointments.[48]

The AWB also recommends that applicants are given clear written reasons why their application for judicial office has been rejected.[49] The Judicial Appointments Division of the LCD does give feedback on applications if requested, but this is not the same as formal written reasons for rejection, matched to the selection criteria and job description, which can form the basis for challenge. In 1997, the Lord Chancellor announced that he was looking into the possibility of appointing an ombudsperson to investigate complaints from failed applicants.[50] Depending on the powers given to the holder of the office, such as access to confidential documents, ability to publish findings and disciplinary powers, this would be a welcome move.

A further measure that the Lord Chancellor has announced is the introduction of an annual report to parliament on judicial appointments, the first report to cover 1998-9 and therefore likely to be published in 2000.[51] How useful this report is will, of course, depend on what it contains. It should include details of the statistics not yet available, such as the applications-appointments ratios for the judicial appointments for which application is required. The AWB has suggested that monitoring of applications should take place to consider not just sex and ethnicity, but also any links, past or present, with Freemasonry[52]; such information could also be disclosed in the annual report. One of the recommendations of the *Without Prejudice?* research was that realistic targets should be set, against which performance can be measured.[53] This recommendation should be adopted with the report setting targets for numbers of women and ethnic minorities in the judiciary. Although the annual report will help improve understanding of procedures, and introduce an element of transparency, of itself it is not enough. Disclosure simply provides the information on which others can base their demands for reform, it does not of itself constitute reform.

Judicial Appointments Commission

Because of the depth of the criticisms of the present appointments process, more progressive suggestions for reform have been advocated over the years. On many occasions in opposition, the now Lord Chancellor, Lord Irvine, committed the Labour Party to a 'radical overhaul of the system for appointing judges', including a judicial appointments system with a strong lay element.[54] Lord Irvine said: 'I do not believe that the present system of appointment guarantees that all deserving candidates will be identified and considered.'[55] Lord Irvine has now reneged on his pledge, and has declared that a judicial appointments commission (JAC) is not a priority and that the system of secret soundings works well.

Reform is, however, needed and a JAC would represent a positive step in the right direction. The role, composition, duties and efficacy of a JAC have been debated at great length, and detailed plans have been promulgated by many organisations.[56] All proposals vary in the exact role given to a JAC, from simply recommending appointments to the Lord Chancellor, to making the appointments itself (the preferred option). In addition, there is great variety as to the possible composition of a JAC, with a lay element common in most proposals.[57] Further scope for variation can be found in the methods of selection to be used. The organisation Justice has advocated the use of selection boards, akin to those used for the civil service; the Law Society maintaining that consultation with senior members of the judiciary should remain a feature of any appointments process.[58] Finally, the criteria to be used in making recommendations or selections is open to debate, and the concerns raised over the nature of the criteria for existing judicial appointments must be borne in mind.

Whatever the format of a JAC, the criticisms of the present system are sufficiently incriminating that its advantages are considerable. First, in varying degrees, it would reduce the role of the executive by taking judicial appointments away from the total control of the Lord Chancellor or Prime Minister. This would help limit the potential for political interference with the appointment of the judiciary, especially regarding senior appointments. Secondly, by including a lay element in the decision making process, less reliance would be placed on the views of the existing judiciary and legal profession, perhaps reducing the potentially discriminatory habit of appointment in one's own mould, what Baroness Helena Kennedy QC has described as 'cloning'.[59] Thirdly, the lay element would meet some of the criticisms of the present 'secret soundings', with less significance being attributed to the culture and traditions of the judiciary, as well as providing the opportunity for more women to participate in the appointments process. Fourthly, it would ensure that the process was transparent, open to all and accountable to the public. This may, in turn, lead to the appointment of a more diverse judiciary. Ultimately, although reform of the judicial appointments process is required, it must also be recognised that the composition of the judiciary depends on the recruitment policies, as well as the culture and ethos, of the legal profession as a whole. The profession as a whole requires reform if women are to become full members of the jurisprudential community.

WILL WOMEN JUDGES MAKE A DIFFERENCE?

Much of the above discussion and debate takes it as axiomatic that there should be more women judges, but why is this? It does not seem that we are simply keen to ensure that women have the opportunity to succeed in their chosen profession, or that we are concerned that drawing the judiciary from a restricted group of individuals denies the existence of highly talented individuals who can contribute to duties of judging – important though these points may be. Is it that, intuitively, we think that it means more than this, in the same way, perhaps, that we think more women MPs will make a difference? But what difference (if any) would it make if there were significant numbers of women judges?

It seems obvious that gender differences will influence judges' decisions. Indeed, it has long been argued, most famously by JAG Griffiths, that social attributes such as class, religion, place and nature of education and race have considerable effects on the nature and content of judgments.[60] It would certainly appear, therefore, that gender could be considered as one of these characteristics which would affect a judge's experiences and therefore judgments. That this may be the case is not just supported by those critiquing the social background of judges, but also finds support in the work of some feminist legal scholars.

There is a strand of feminist thinking which considers that women have different behavioural characteristics to men that are inherent in their sex. The most famous proponent of this view is the psychologist Carol Gilligan whose book, *In a Different Voice*, remains the reference point for this debate and still excites great controversy.[61] Carol Gilligan argues that women are more connected to others, in contrast to the autonomy and independence of men, and are therefore better at relationships, are driven by an ethic of care and responsibility, rather than a male ethic of rights, and have greater empathy with others. In terms of judging, the argument is made that as women have such different characteristics, more women judges will lead to a radically different law and legal system as women use their skills and perspectives to interpret the law in a different manner to men.[62]

With this hypothesis in mind, numerous studies in the US have been carried out seeking to discover any gender differences in judgments.[63] Despite the dubiety of some of the research methods – how do you decide whether a judgment was given on the basis of gender or some other factor? – it has been suggested that Justice Sandra Day O'Connor of the US Supreme Court dispenses justice with a uniquely feminine jurisprudence which emphasises 'connection, subjectivity, and responsibility', rather than 'autonomy, objectivity and rights'.[64] However, it has to be said that many others have found 'few, if any, consistent, and statistically demonstrable differences among judges based on gender'.[65]

On many levels, these arguments are problematic. Australian law professor Regina Graycar has argued that asking whether women will make a difference raises the question, different to what?[66] The answer is different to men, which therefore assumes that men and men judging are the norm. When Lord Justice Otton told the 1996 Woman Lawyer Conference that women barristers bring different perspectives to cases, he meant different *to men*, reinforcing the view that the perspectives of men were normal, women the odd ones out. This is most clearly exemplified by cases in which white women, black women and black men have been judges, and in respect of which litigants have brought cases alleging 'bias'.[67] A classic of this type is the case in which a Melbourne solicitor sought judicial review of a planning tribunal decision on the grounds that a tribunal member, five months pregnant at the time of her decision, 'suffered from the well-known medical condition "placidity" which detracts significantly from the intellectual competence of all mothers-to-be'.[68] Needless to say, there are no cases in which a challenge has been made to a judge, or other decision maker, on the basis that he was white, male and this had constrained his decision making by adopting a white male perspective.[69] In addition, this view of women as having these 'difference characteristics' is too deterministic. It assumes that 'women' is a homogenous category in which all women think the same, have the same experiences and will act in the same way.

In 1991 Sandra Day O'Connor gave a lecture in which she considered this question of whether women judges will make a difference.[70] Justice O'Connor traced the historical barriers to women entering the legal profession, and then commented that just as women are beginning to achieve successes, they are faced with arguments, albeit feminist in intention, about how they are so very different from men lawyers and judges. She argues that these debates 'recall the old myths we have struggled to put behind us'.[71] The one difference between women and men lawyers and judges, in the view of Justice O'Connor, is that women professionals still have primary responsibility for children and housekeeping.[72] As a result of this, many women face a glass ceiling, or are sidelined into 'mommy track' jobs. Thus, women need to be treated differently to the extent that account must be taken of pregnancy, in terms of leave and pay, but that does not mean that *all* women are different from *all* men in the manner in which they perform their professional obligations.

In addition, it has to be recognised that we all bring our own perspectives, experiences and views to bear on our processes of reaching decisions. Simone de Beauvoir said that 'it is impossible to approach any human problem with a mind free from bias'.[73] Regina Graycar has explored how judges resort to what they consider to be 'common sense' when making judgments. She gives many examples such as the infamous comment by the Australian Judge Bland when he remarked that: 'it does happen, in the common experience of those who have been in the law as long as I have anyway, that no often subsequently means yes'. Similarly, in the context of a damages claim, a judge commented regarding remuneration for domestic work that 'the sharing of domestic burdens with the wife is expected of the husband, even where his wife is perfectly healthy'.[74] The judges use their knowledge and understanding of the world to construct a common sense approach to factual questions before them. Thus, the criticisms of JAG Griffiths are largely bound up in the fact that judges bring a particularly establishment view to their judging, a view which is seen as apolitical. Inevitably, therefore, every 'decision-maker who walks into a courtroom to hear a case is armed not only with the relevant legal texts, but with a set of values, experiences and assumptions that are thoroughly embedded'.[75]

Accordingly, a judiciary which is drawn from a wider range of backgrounds, sexes and ethnicities will bring different experiences to bear on their judgments. Women, especially as they come through the legal system at present, will often have quite different experiences to their male colleagues, whilst others will have markedly similar approaches to male colleagues. As Canadian Supreme Court judge, Justice Bertha Wilson points out, it would be a 'Pyrrhic victory for women and the justice system as a whole if changes in the law come only through the efforts of women lawyers and women judges'.[76] Although she concludes that if 'women lawyers and women judges through their differing perspectives on life can bring a new humanity to bear on the decision-making process, perhaps they *will* make a difference'.[77] In a similar vein, the influential Hansard Society report *Women at the Top* considered that the role of women in the judiciary was crucial as the 'decisions taken by judges … have far-reaching consequences for all members of society, and the under-representation of women among their ranks limits the quality and vision of these decisions'.[78]

In addition, the existence of more women judges would mean a greater number of mentors and role models for aspiring women lawyers. This would be especially important if the present appointments system were to continue. For example, Hazel Williamson QC has suggested that women exhibit a 'diffidence about putting themselves forward as worthy candidates' for the judiciary.[79] Similarly, Cherie Booth QC has said: 'Men colleagues automatically assume they will take silk or get a judicial appointment. They have much more faith in their ability to glide effortlessly from one career step to another. Women are much more hesitant about whether they can do that.'[80] The *Without Prejudice?* research suggested that the secrecy of the selection system, the lack of information about criteria, the possible experience of discrimination and patronising treatment from male judges who will determine selection and a professional culture which condones stereotypical and discriminatory remarks, all conspire against women considering themselves for judicial office.[81] A greater number of senior women, particularly women judges, could help, in an informal mentoring capacity, to reduce some of these negative impulses.

In another way, more women judges will make a difference. This chapter has considered how women are often seen as less authoritative than men, and therefore less suited to judicial office. An increase in the number of women judges will begin to reduce the survival of such views. This in turn will have an effect on women in the profession and in law schools, making judicial appointment a more common aspiration. Thus, as US law professor Suzanna Sherry has argued, the mere fact that women are judges has an educative function: 'it helps to shatter stereotypes about the role of women in society that are held by male judges and lawyers, as well as by litigants, jurors and witnesses'.[82] Thus, for women law students, it will be normal that judges are women, validating their studies and their role in the law. Women and men in the profession will also become accustomed to seeing women in authoritative legal roles. Finally, it has been suggested that women advocates appearing before women judges, as opposed to some men judges, find the experience liberating, as they are not seen as 'out of place' and having to justify their presence.[83]

CONCLUSIONS

It is undoubtedly the case that progress towards a more diverse and open judiciary is being made. Reforms are being implemented, and more women and lawyers from ethnic minorities are being appointed. This must raise the public's confidence in the legal process. However, the UK lags far behind some jurisdictions, for example France, Germany, Italy and the Netherlands, where women are being appointed judges in equal numbers to men.[84] Furthermore, in North America, where more open and publicly accountable appointments procedures have been the norm for many years, women are appointed to the Bench in considerable numbers.[85] Although the composition of the judiciary has excited debate for many years, it will become increasingly important as the judiciary takes on the role of interpreting the European Convention on Human Rights. And, if indeed there is now 'a culture of judicial assertiveness to compensate for, and in places repair, dsyfunctions in the democratic process', as Mr Justice Sedley has suggested[86], it is imperative that the judiciary is more representative of society as a whole.

PERSONAL TESTIMONIES OF WOMEN JUDGES

The Honourable Mrs Justice Hale DBE

Mrs Justice Hale first made her name as an academic, then as a Law Commissioner and is now one of seven women High Court judges. She relates below how her career has developed and gives her views on women in the law ...

'My career has been unusual, even unique, but it is hard to say whether this is in any way due to my being a woman. I only thought of law when the headmistress of the small rural girls' high school (which my sisters and I attended) was afraid that I was not enough of a natural historian to get into Oxbridge. Instead of protesting that law was a man's subject, she gave me every encouragement. The outcome was an exhibition to Girton College, Cambridge.

We were told the usual tales about how hard it was to make a start and earn a living at the Bar. In those days, it was wise to join an Inn and begin eating dinners as soon as possible. I could not afford it. With no experience, no connections and no evidence of later academic success, it is hardly surprising that the Inn refused me a scholarship. During the long vacations I worked for our local market place solicitor, who kindly (?) said I should aim higher, and for a top City firm of solicitors, who offered me articles. But London and exclusively commercial clients did not appeal.

Demand for law teachers was expanding in the 'post-Robbins' era. This was a chance to gain some experience, earn a little money, and qualify as a barrister by home study as one could in those days. Manchester not only had well-known professors, but also encouraged me to qualify and practise part-time.

But by 1972 it was one or the other: advancement at the Bar was prevented by not being able to take long cases during term time, advancement in the University was prevented by not having so much time to research and write. I chose the latter: my first husband was at the Manchester Bar and we wanted to have a family. One daughter and four books (on *Mental Health Law*, *Parents and Children*, *The Family, Law and Society* and *Women and the Law*) followed.

The first book led in 1979 to membership of the mental health review tribunals. Being a founder editor of the Journal of Social Welfare Law led in 1980 to membership of the Council on Tribunals. Together these must have led in 1982 to being asked to sit as a part-time judge and in 1984 to membership of the Law Commission. The Family Law team played a large part in the Reports leading to the Children Act 1989 and to the Family Law Act 1996. The Commission is a full-time job, but could be combined with sitting as a part-time judge for at least four weeks each year in crown courts and county courts in the London area. In 1989 I became a Queen's Counsel under the then category of 'employed barrister' and began to sit as a Deputy High Court Judge in the Family Division.

In 1994, I was appointed a High Court Judge and assigned to the Family Division. Some High Court judges have experience as academic lawyers, but as far as anyone knows, I am the first to have made a name as an academic and Law Commissioner, rather than a legal practitioner. My career is testimony to the sort of flexible approach that enables women to have real equality of opportunity in otherwise male-dominated worlds. But one swallow does not make a summer. One reason why women are still seriously under-represented in the judiciary is the assumption that the best qualification for the best judicial appointments is long and successful practice in the courts.

My impression is that the Bar and Judiciary are much less female friendly than are education and the public service. The latter have been formally committed to appointment purely on merit for a very long time. They now have equal opportunity policies as well. As large organisations rather than loose associations of individual practitioners it is easier for them to do so. They have been able to adapt into a genuine two sex profession.

The Bar and Judiciary still give the impression that they are a man's profession which women are allowed to join. The men are usually kind to us, but they have made few changes to the way they behave in order to accommodate us. The judiciary is particularly slow to change, which is hardly surprising. There are still very few women members. The decision-making structure is a combination of fierce judicial independence on some matters and deference to authority or seniority in others. There is little tradition of consultation and collectivity.

No-one worth her salt would allow herself to be put off by any of this. There are many different ways to succeed in the law: diligence, determination and a degree of aptitude are what it takes, but being in the right place at the right time is a great help.'

Dame Rosalyn Higgins QC

Dame Rosalyn Higgins became the first woman judge of the International Court of Justice when she was appointed in 1995. She describes below the nature of her work and her career path ...

'The International Court of Justice is the judicial arm of the United Nations. It has its seat at the Hague and is the legal successor to the Permanent Court of International Justice, established in 1922. It decides cases brought between states and gives advisory opinions to those UN organs and international organisations competent under the UN Charter to ask for them.

While it thus has a contentious and an advisory role, litigation between states forms by far the larger part of its docket. Recent and current cases include disputes between Hungary and Slovakia; Bahrain and Qatar; Botswana and Namibia; and Spain and Canada.

The Judges of the Court are from the various corners of the world and are elected by the UN Security Council and General Assembly. I arrived there in July 1995 as the British Judge and as the first woman judge to sit in either the Permanent Court or the present International Court.

Working with my 14 fellow judges feels no different from much of my previous working life, where I have always operated in predominantly male surroundings. Equally, then as now, I have never felt in my work that my gender makes the remotest difference. At the Bar, where I practised as a silk in international and petroleum law, and at LSE, where I was Professor of International Law from 1981-1995, I was in a numerical minority as a woman. Beyond whatever implications that fact may bear, it has been an irrelevance. I have felt, all my professional life, that I was simply working with colleagues, where gender was irrelevant (as was religion and colour). I appreciate that this experience may be exceptionally fortunate.

I went to Burlington Grammar School from 1943–1955, belonging to that generation whose parents had left school at age 13 and who therefore regarded grammar schools as an exceptional opportunity for their children. No-one in my family was a lawyer nor did I know any lawyers. A school teacher suggested that I might do quite well at the subject. I got a scholarship to Cambridge and entered a new world.

It was suggested to me that a woman could not expect to succeed in international law and that a career in a different field of law might be more appropriate. In the event, I found no barriers at all in the world of international law.

The very nature of international legal work – frequent travel to far-flung places, intensive work often in evenings and weekends, the need to combine practice with teaching and hence the burden of doing two full-time jobs – is perhaps particularly hard for women. One needs an extraordinarily tolerant husband, who will simply regard these things as 'normal' and get on with his own career (and this I have had). And one needs very stable arrangements to ensure the security of any children. I realise, too, that my ability to put that in hand was a privileged option. In any event, eschewing 'live-in-nannies' or au pairs, or boarding school, but with good daily help, I have never felt an 'absent mum' – and my children confirm that they never saw it that way, either.

Good fortune has conspired to make me feel that one can indeed 'have it all' – senior level career and a happy family life. I appreciate that for many others – facing the problem of a career structure in a city law firm, or without a supportive husband who considers this way of life normal, or in severe financial constraints – this is not a reality they recognise. But as I see my former students, women as well as men, making important careers for themselves and assuming international positions of importance, I am very optimistic about the future of women in international law.'

District Judge Gillian Marley

Judge Gillian Marley started off her career as a barrister. She then became a solicitor and is now a District Judge. She describes below the steps she has taken in her career and the nature of her work as a District Judge ...

'My interest in law dates from my school days. At the age of sixteen, on a school trip to London I slipped away to the Old Bailey and was inspired by a criminal trial whilst the others explored the Tower of London. At my Convent School in Devon the nuns did not approve of my ambition to become a lawyer as they felt that the law was a rather disreputable profession; a few years ago when they learned that I had become a District Judge they appeared to forget their former anxiety.

I wanted to study law at university but not having done any law previously and having little idea what it would be like I opted for a Joint Honours degree in Law and Economics at Durham University. I enjoyed the Law but not the Economics and after the first year transferred to Law and Sociology. I was the only woman in my year studying law; no problems there; and I graduated with a Joint Honours degree.

After graduating I obtained a job as a law lecturer at a polytechnic in South London teaching law to Accountancy students. The first year was a challenge but I found the second year repetitive and decided that teaching was not for me. The only aspect I had really enjoyed was taking the students to court. I sought advice from my law tutor at Durham University who suggested I should study for the Bar Exams. I joined Lincoln's Inn, ate the necessary dinners, passed the exams and was called to the Bar.

My pupillage was spent in London. It was an exciting time, spent in a variety of courts from Gravesend Magistrates Court to the House of Lords. On completing my pupillage I found there was no prospect of my remaining in the same chambers; they had never allowed a female to join chambers as a tenant and did not intend to alter that practice. I applied to a new set of chambers which had recently opened in Middlesbrough, where I was accepted.

There had not been any women barristers on Teesside previously and it took a while to become established. Eventually local solicitors realised that there could be advantages in instructing a woman barrister and some of their clients (especially women) preferred to be represented by a woman. I thoroughly enjoyed the work; a combination of civil, criminal and family law which is familiar to many junior barristers.

By this time I was married and after eight years at the Bar I had two sons within fourteen months. I had intended to continue to practice on a part-time basis but found it too difficult to combine this with motherhood so I decided to take a career break for five years to enable me to look after my sons whilst they were small. I would not have missed those years and I feel it is unfortunate that so many young professional women are unable to enjoy their children's formative years. I continued to do a little part-time law lecturing at a local college.

When the children were both at school, I was ready to return to a career (on a part-time basis initially). I applied to be voluntarily disbarred, took two additional exams and was enrolled as a solicitor. I did this because I thought that part-time work as a solicitor would be more compatible with the demands of a young family than attempting to resume a career at the Bar, with its irregular working hours, unpredictability and frequent travel.

I obtained a position with a local firm of solicitors, dealing with litigation and family cases, including advocacy in the lower courts. It was interesting to see a case through from start to finish, whereas the barrister often sees only the day in court. I was concerned to find that women solicitors were often paid less than their male colleagues and there were not many women partners.

After seven years in practice as a solicitor I was appointed as a Deputy District Judge. I was asked to attend an interview at the House of Lords; I then went on a residential training course and 'sat in' with an experienced District Judge for several days. I was then qualified to sit as a Deputy District Judge for about thirty days a year whilst continuing in practice as a solicitor.

I found this new type of work extremely interesting and very varied and after three years as a Deputy I applied successfully for a full time appointment. I was appointed initially to a court about seventy miles from home. I travelled each day until I obtained a transfer to a court nearer to home.

I can say that as a District Judge I have (for the first time in my career) experienced no discrimination on grounds of gender. The LCD has an equal opportunities policy and women are encouraged to apply for judicial appointments. The work of a District Judge is fascinating. It includes a wide range of civil work from insolvency to commercial to general litigation, family cases and the Small Claims Court. Some District Judges are trained and nominated for child care work. Our jurisdiction has increased considerably in recent years. I would encourage suitably qualified women to apply to sit as deputy district judges.

I am grateful to my family (especially my husband) for their support and encouragement in my career and my Christian faith which helps me to keep it all in perspective.'

The Honourable Mrs Justice Arden DBE

Mrs Justice Arden was called to the Bar at a time when there were few women barristers, especially in the commercial field. This did not put off Mrs Justice Arden who became the first woman appointed to the Chancery Division of the High Court and the first woman to be appointed Chairman of the Law Commission. She is also President of the Association of Women Barristers. She describes below her path to the law and offers some advice for women lawyers ...

'When I was at school in the 1960s we did not have careers' departments and we were expected to find out for ourselves what careers we would enjoy. So I followed the best example that anyone could have: my parents. My father was a solicitor and my mother had worked with him prior to getting married. My father had learnt about the law from his father and had an instinctive feel for legal reasoning and the high standards attached to being a practising lawyer. I worked in his office sometimes in the school holidays. I liked what I saw of the law and decided I too would be a lawyer. By a stroke of great good fortune, I obtained a place to read law at Cambridge and indeed stayed there for four years.

In my final term, I decided that it seemed that barristers had all the fun of helping to make new law and that I would become one too. My father was initially sad at my decision. The organisation of the profession is not suitable for women he said – rightly at the time. But in my home town, there were then some very distinguished and able women barristers, including Rose Heilbron QC, a truly great woman lawyer, and advocate who subsequently became a High Court Judge, and so my father knew it *could* be done.

I then had a splendid year studying in the United States at Harvard Law School, later passing the Bar finals (now the BVC) and began looking for a pupillage. Through the help of a friend of my father, I was found an extremely good pupil master. There was very short hiatus between pupillage and finding a set in Chambers. I wanted to practise in an area of commercial law and those were the days when people turned you down with such phrases as 'we have a woman already' or 'commercial solicitors would never instruct a woman'. Those days are over.

Then followed twenty-two years of practice at the Bar, seven as a Queen's Counsel. I married another barrister in a similar field to my own and we had three children. That certainly made demands on the skills that we women are supposed to have for multi-tasking. The organisation and structure of the legal profession have changed since women joined, but the pressures and demands of professional and family life on women are still substantial.

My husband and I were one of the small number of barrister couples. We then became one of the few pairs of QCs and now we are the only married couple who are both High Court Judges. I was the first woman to be appointed a Judge of the Chancery Division of the High Court. In 1996, I was appointed Chairman of the Law Commission.

I would offer the following general advice to women lawyers. First, be yourself. Think about what you would like to achieve in your personal and professional life and see how you can get closest to it. Don't push yourself out completely just to have a successful career, important though that is. Second, plan it. Find out what is involved for as far as you can into the future. Try to see the upsides and downsides of the various courses open to you. Third, enjoy the law. The law is immensely interesting and stimulating and it has many different facets: there are many different fields of law and you do not have to be a lawyer in private practice to use your law: you could go into industry or government to mention just two of the other options. Decide whether it is helping people or solving their problems that you do best and that will help you choose a field of law that will bring out your gifts if you decide to specialise. Fourth, when you start your career, keep in touch with other women lawyers. Many of the problems that you face will be problems that other women have or have had too, and by discussing them together you will often find that what you took for a real difficulty is something that can easily be overcome.'

NOTES

1 TMS Consultants, *Without Prejudice? Sex Equality at the Bar and in the Judiciary*, Bar Council and LCD, 1992, para 48(I).
2 Quoted in Barbara Hewson, 'Why women get a raw deal in the courts', *The Times*, 17 September 1996.
3 Margaret Thornton, *Dissonance and Distrust – Women in the Legal Profession*, Oxford University Press: Oxford, 1996, p 291.
4 Http: //www.open.gov.uk/lcd/judicial/women.htm.
5 Figures as of 1 February 1998 and quoted from http://www.open.gov.uk/lcd/judicial/womapp.htm. 1989 statistics taken from Hansard Society, *Women at the Top*, Hansard Society: London, 1990, p 45.
6 See EOC evidence to the Home Affairs Committee, Third Report of Session 1995–96, Volume II, *Minutes of Evidence and Appendices*, p 218.
7 Quoted in 145 (1995) *New Law Journal* 514.
8 Sally Hughes, *The Circuit Bench – A Woman's Place?*, Law Society: London, 1991.
9 Supra note 7.
10 1997 statistics provided by the LCD. For 1994, see EOC, supra note 6, p 218. The data may be incomplete as applicants for judicial office have only been asked their ethnic origin since 1991.
11 Lord Chancellor, Speech to Minority Lawyers' Conference, 29 November 1997, reproduced in full at: http://www.open.gov.uk/lcd/speeches/minlaw.htm.

12 LCD Press Notice, 'Judicial Appointments Commission', 125/97, 23 June 1997.
13 Speech to AWB reproduced at http://www.open.gov.uk/lcd/judicial/women.htm.
14 Ibid.
15 Although this process is under review.
16 LCD Press Notice, 'First advertisements for High Court Judges', 43/98, 23 February 1998.
17 See the LCD web site: http://www.open.gov.uk/lcd/lcdhome.htm.
18 Joshua Rozenburg, *The Search for Justice*, Sceptre: London, 1994, pp 9–10. It was also reported that the former Lord Chancellor, Lord Mackay, had to fight off a right wing challenge, led by Michael Howard, to the appointment of the 'liberal' judges Lords Bingham and Woolf to, respectively, Chief Justice and Master of the Rolls: *The Observer*, 26 May 1996.
19 Joshua Rozenburg, ibid p 9.
20 Supra note 6, p 212 and AWB evidence to Home Affairs Committee, Third Report of Session 1995–96, Volume II, *Minutes of Evidence and Appendices*, p 193. For further information on the AWB, see appendices five and six.
21 As reported in the House of Commons Home Affairs Committee, Session 1995–96, Third Report, *Judicial Appointments Procedures*, Volume I, paras 150 and 128 respectively.
22 Supra note 11.
23 Supra note 1.
24 Quoted in 'The Second Woman Lawyer Conference', *Counsel*, May/June 1996, p 8.
25 Quoted in Helena Kennedy, *Eve was Framed*, Chatto & Windus: London, 1992, p 267.
26 Josephine Hayes, 'Appointment by invitation', 147 (1997) *New Law Journal* 520–522, p 521.
27 Ibid.
28 Supra note 20, p 193.
29 Ibid p 194.
30 Home Affairs Committee, Third Report Session 1996–97, *Freemasonry in the Police and the Judiciary*, para 18. For the first time, the advertisement for the High Court bench in early 1998 gave a warning that an appointee may be required to disclose, as a condition of their appointment, whether they belong to the Freemasons, or to notify the Lord Chancellor if they subsequently join. This is in the light of a positive Government response to the recommendations of Parliament's Home Affairs Committee that there should be disclosure of membership of the Freemasons.
31 Supra note 6, p 211.
32 Ibid.
33 Reported in 'Law Society clashes with the Lord Chancellor', 141 (1992) *New Law Journal* 1062.
34 Ibid.
35 Reported in Glenda Cooper, 'Lord Irvine, and a spot of bother with women', *The Independent*, 21 February 1998.
36 Reported in 'Partner leads backlash against Govt appointment procedures', *The Lawyer*, 10 February 1998.

37   Frances Gibb, 'Inquiry into "secret" law jobs', *The Times*, 26 February 1998. In a similar vein, two actions have been brought against the LCD on the basis that the Lord Chancellor appointed an advisor (a male friend) without the post being advertised and there being a proper selection procedure. Solicitor Jane Coker has brought a sex discrimination claim, and Martha Osamor is alleging sexual and racial discrimination.

38   Supra note 20, para 119.

39   Ibid.

40   Specifications are detailed on the LCD web site, supra note 17, and information packs on each appointment are available from the Judicial Appointments Division of the LCD, contact details of which can be found in appendix five.

41   Quoted in Clare Dyer, 'Lord Mackay urges more women to apply for silk', *The Guardian*, 10 April 1995.

42   Ibid and supra note 6, p 211.

43   See chapter six under the heading 'Taking silk' and chapter four under the heading 'A baby bar?'.

44   As suggested by Helena Kennedy, supra note 25, p 268.

45   Supra note 6, p 211.

46   Supra note 11.

47   Supra note 6, p 211.

48   Supra note 20, p 192.

49   Ibid p 191.

50   *The Lawyer*, 14 October 1997.

51   LCD Press Notice, 'Lord Chancellor announces new measures for judicial appointments', 220/97, 23 June 1997.

52   Supra note 20, p 194.

53   Supra note 1, pp 26–27.

54   For just one example, see Lord Irvine quoted in Clare Dyer, 'Women QCs "enjoy reverse bias"', *The Guardian*, 30 September 1996.

55   Quoted in Frances Gibb, 'Labour promises to bear down hard on legal aid costs', *The Times*, 30 September 1996. See also Lord Irvine, 'The Legal System and Law Reform Under Labour', in David Bean (ed), *Law Reform in All*, Blackstone Press: London, 1996, p 20.

56   For example, Justice, *The Judiciary in England and Wales*, Justice: London, 1992.

57   Supra note 21, paras 133–136.

58   Ibid, para 135-6.

59   Supra note 24, p 267.

60   JAG Griffith, *The Politics of the Judiciary*, (5th edn), Fontana Press, 1997. See also, 'Judging from on high', 86 (7) (1997) *Labour Research* 13–15.

61   Carol Gilligan, *In a Different Voice*, Harvard University Press: Harvard, 1982.

62   See Carrie Menkel-Meadow, 'Portia in a different voice: speculations on a woman's lawyering process', 1 (1985) *Berkeley Women's Law Journal* 39–63.

63   See, for example, Jilda Aliotta, 'Justice O'Connor and the equal protection clause: a feminine voice?', 78 (1995) *Judicature* 232–235.

64   Suzanna Sherry, 'Civil virtue and the feminine voice of constitutional adjudication' 72 (1986) *Vanderbilt Law Review* 543–615, p 593.

65   *Supra* note 63, p 235.

66 Regina Graycar, 'The Gender of Judgments: An Introduction', in Margaret Thornton (ed), *Public and Private – Feminist Legal Debates*, Oxford University Press: Oxford, 1995, pp 262–282.

67 Ibid p 267.

68 See Bronwyn Naylor, 'Pregnant Tribunals', 14 (1989) *Legal Service Bulletin* 41.

69 Supra note 66, p 267.

70 Sandra Day O'Connor, 'Portia's Progress', 66 (1991) *New York University Law Review* 1546–1558.

71 Ibid p 1553.

72 Ibid. A point echoed by Mrs Justice Arden at the 1998 Woman Lawyer Conference: see Frances Gibb, 'Sexual bias is still a problem, says a Bar pupil', *The Times*, 27 April 1998.

73 Simone de Beauvoir, *The Second Sex*, translated by H M Parshley, Penguin Books: London, 1953, p xxxii.

74 Supra note 66, p 270.

75 Rosalie Abella, 'The Concept of an Independent Judiciary', in S Martin and K Mahoney (eds), *Equality and Judicial Neutrality*, Carswell: Toronto, 1987, pp 8–9.

76 Justice Bertha Wilson, 'Will women judges really make a difference?', 28 (1990) *Osgoode Hall Law Journal* 507–522, p 516.

77 Ibid p 522, emphasis in the original.

78 Hansard Society, *Women at the Top*, Hansard Society: London, 1989, p 44.

79 Hazel Williamson QC, 'Networking and gaining promotion', Handout, 1995 Woman Lawyer Conference.

80 Quoted in 'The Second Woman Lawyer Conference', *Counsel*, May/June 1996, p 8.

81 Supra note 1, p 24.

82 Suzanna Sherry, 'The Gender of Judges', 4 (1986) *Law and Inequality* 159, p 160.

83 Dianne Martin, 'Have women judges really made a difference?', 6 (1986) *Lawyers Weekly* 5, p 5.

84 Cheryl Thomas, 'Judicial Appointments in Continental Europe', Research Series No 6/97, LCD, 1997, p 22.

85 See for example: Kate Malleson, 'The Use of Judicial Appointments Commissions: A Review of the US and Canadian Models', Research Series No 6/97, LCD, 1997 and Ruth Bader Ginsburg and Laura Brill, 'Women in the Federal Judiciary', 64 (1995) *Fordham Law Review* 281–290.

86 Stephen Sedley, 'Human Rights: a twenty-first century agenda', (1995) *Public Law* 386, p 388.

Appendix 1
Types of law degrees

The law degrees offered at universities vary considerably in both content and structure. This appendix outlines the principal structural differences. Reference should be made to the UCAS handbook and other similar texts for the most up to date information.

Traditional three year law degree (LLB)

A full-time three year law degree is offered by most universities. This degree is likely to cover the seven foundational law subjects which exempt a student from the first stage of professional legal training, as well as some options. Some degrees will allow students to take a number of course options outside the law department which can be a good opportunity to continue an interest in another subject.

Four year sandwich degree

A further option is a four year course where either the third year, or a number of periods throughout the degree, is spent on a work placement in a legal establishment. Sandwich degrees are available at the universities of Brunel, Bournemouth, Nottingham Trent and Sheffield Hallam University. Jane Hyndman, a solicitor with Live TV, studied law at Brunel University, taking their four year sandwich course. 'This enabled me to spend three periods of six months on placement in each of the first three years of the degree', says Jane. 'These were spent at the DPP in Belfast, a small high street general practice in central London, and in the legal department of the Central Electricity Generating Board. The nature of the work was quite different on each placement and this enabled me to gain a wide variety of experiences at a very early stage in my career and enabled me to make an extremely well informed choice about the type of firm in which I wished to train.'

Mixed/joint degree

Many universities also offer a joint or mixed degree. This type of degree combines the study of law with another subject, for example, a language, history, sociology or accounting. Most degrees of this type enable the student to study those law subjects necessary to achieve exemption from the first stage of vocational training for the legal profession. District Judge Gillian Marley recalls that she wanted to study law at university but, having little idea really what law was about, she opted for a joint honours degree in Law and Economics at the University of Durham. Judge Marley enjoyed the law, but not the economics, and after the first year switched to another joint degree, this time Law and Sociology.

Two year degree

A two year degree is currently offered by the University of Buckingham. The course covers many of the same subjects available at other universities but teaches them over four terms, as opposed to the three terms structure of other universities, hence the shorter completion time. The downside of study at Buckingham has always been that students have to pay their own fees as it is a private university. It may be that with the advent of tuition fees being charged for all university education, the option of the University of Buckingham may become more popular.

Four year LLB European legal studies

Many universities now offer four year degrees which involve one year abroad at a European university. This option has become increasingly popular and is possible because of funding from the European Union. Each university which offers this degree has partnerships with different European universities. Competency in a particular language is obviously a prerequisite for most of these courses. Some law schools offer the option to transfer to this degree having started a standard law degree and some universities have partnerships with European universities which teach in English.

Four year 'exempting' degree

The University of Northumbria at Newcastle offers a four year degree which combines academic studies with the learning of the legal skills necessary to provide exemption from the vocational stage of training for the legal profession (the LPC or BVC). Thus, whereas becoming a solicitor or barrister requires one further year of study after graduating with a qualifying law degree, the Northumbria degree includes that extra year within the course. The added attraction is that the fees for the four years are paid by government (although from 1998 a contribution to those fees will be required), making it much cheaper in the long run. However, students must decide before they embark on their degree course whether they wish to take the degree leading to qualification as a solicitor or barrister.

Part-time law degree

The majority of 'new' universities offer part-time law degrees, as does Birkbeck College and the University of Hull. The growth in part-time degrees has played an important role in the general expansion of higher education and the number of law students.

Open University law degree and distance learning

The Open University (OU) began offering law degrees from early 1998, attracting 900 students for its first course. A degree from the OU can count as a qualifying law degree (and therefore proving exemption from the first year of vocational training for the legal profession) if the requisite courses are completed within six years.

Distance learning, other than with the OU, is available from a number of institutions, including Holborn College, which is an associate college of Wolverhampton University, Manchester Metropolitan University, Nottingham Trent University and the external London University LLB.

Four year 'English law' degree at University of Dundee

The University of Dundee offers a four year law degree which is a qualifying law degree for the purposes of exemption from the first part of the vocational stage of training for the English legal profession. It is also possible to study the necessary subjects to qualify for the Scottish legal profession.

Scottish law degrees

All Scottish universities offer an LLB degree in Scots law, some of which are available part-time. These degrees are based on Scots law which is quite different from English law. For a student wishing to pursue a career in the English legal profession, exemption from the CPE may be given for some subjects studied, such as European Union law, but further study would be required for other courses such as criminal law.

Appendix 2

UK university law schools offering a course considering the law from a gender/feminist perspective

University of Birmingham
Women and the Criminal System

University of Bristol
Gender and the Law

University of Brunel
Law, State and Gender

University of Buckingham
Sex, Gender and the Legal Process

University of Cambridge
Women and the Law

University of Coventry
Women and the Law

University of Dundee
Gender and the Law

University of East London
Women and the Law

University of Edinburgh
Gender and Justice

University of Keele
Women and the Law

University of Kent
Feminist Perspectives of Law*

University of Lancaster
Gender and the Law

*not run for past two/three years

University of Liverpool
Law and the Sexes

University of Manchester
Gender and the Law

Manchester Metropolitan University
Gender and the Law

Middlesex University
Women and the Law

Nottingham Trent University
Sexuality and the Law

Oxford Brookes University
Sexuality and the Law

Queen's University
Gender and the Law*

University of Sheffield
Women and Crime

South Bank University
Women and the Law

Staffordshire University
Feminism, Post-structuralism and the Law

University of Warwick
Women and the Law

University of West of England
Gender and the Law

University of Wolverhampton
Women and the Law

University of Westminster
Women and Law

No information is available in respect of Anglia Polytechnic University, Aston University, University of East Anglia, Luton University, Oxford University and Queen Mary and Westfield College. For details of the methodology of the survey, see appendix 3.

Number of academic staff, and availability of undergraduate and postgraduate courses considering the law from a gender/feminist perspective, in UK law schools

University of Aberdeen, Faculty of Law

	Men	Women	Total	% Women	UK % Women
Professoriat	10	0	10	0	14
Senior	3	2	5	40	40
Lecturers: permanent	8	3	11	27	44
Lecturers: fixed term	0	0	0	0	49
Total staff	21	5	26	19	39

undergraduate course/s considering the law from a gender/feminist perspective: **none**

undergraduate course/s considering the law from a gender/feminist perspective in parts: **none**

postgraduate degree/course/s considering the law from a gender/feminist perspective: **none**

University of Abertay Dundee, School of Accountancy and Law

	Men	Women	Total	% Women	UK % Women
Professoriat	0	0	0	0	14
Senior	3	0	3	0	40
Lecturers: permanent	4	3	7	43	44
Lecturers: fixed term	1	1	2	50	49
Total staff	8	4	12	33	39

undergraduate course/s considering the law from a gender/feminist perspective: **none**

undergraduate course/s considering the law from a gender/feminist perspective in parts: **none**

postgraduate degree/course/s considering the law from a gender/feminist perspective: **none**

University of Wales, Aberystwyth, Faculty of Law

	Men	Women	Total	% Women	UK % Women
Professoriat	2	2	4	50	14
Senior	3	2	5	40	40
Lecturers: permanent	12	9	21	43	44
Lecturers: fixed term	0	0	0	0	49
Total staff	17	13	30	43	39

undergraduate course/s considering the law from a gender/feminist perspective: **none**

undergraduate course/s considering the law from a gender/feminist perspective in parts: **none**

postgraduate degree/course/s considering the law from a gender/feminist perspective: **none**

Birkbeck College, University of London, Department of Law

	Men	Women	Total	% Women	UK % Women
Professoriate	2	1	3	33	14
Senior	6	2	8	25	40
Lecturers: permanent	0	0	0	0	44
Lecturers: fixed term	1	0	1	0	49
Total staff	9	3	12	25	39

undergraduate course/s considering the law from a gender/feminist perspective: **none**

undergraduate courses considering the law from a gender/feminist perspective in parts:

gender issues addressed in many courses including:
Contract Law
Criminal Law
Criminal Justice
Jurisprudence
Family Law
Tort Law
Human Rights

The Department seeks to include gender issues in all courses

postgraduate degree/course/s considering the law from a gender/feminist perspective: **none**

University of Birmingham, Faculty of Law

	Men	Women	Total	% Women	UK % Women
Professoriat	6	1	7	14	14
Senior	11	1	12	8	40
Lecturers: permanent	10	6	16	38	44
Lecturers: fixed term	1	1	2	50	49
Total staff	28	9	37	24	39

undergraduate course considering the law from a gender/feminist perspective:
Women and the Criminal Justice System

undergraduate courses considering the law from a gender/feminist perspective in parts:
Jurisprudence
Juvenile Justice
Criminal Law and Justice

postgraduate degree/course/s considering the law from a gender/feminist perspective: **none**

University of Bournemouth, School of Finance and Law

	Men	Women	Total	% Women	UK % Women
Professoriat	1	0	1	0	14
Senior	9	6	15	40	40
Lecturers: permanent	0	0	0	0	44
Lecturers: fixed term	0	1	1	100	49
Total staff	10	7	17	41	39

undergraduate course/s considering the law from a gender/feminist perspective: **none**

undergraduate course/s considering the law from a gender/feminist perspective in parts: **none**

postgraduate degree/course/s considering the law from a gender/feminist perspective: **none**

University of Bristol, Department of Law

	Men	Women	Total	% Women	UK% Women
Professoriat	10	2	12	**17**	**14**
Senior	4	1	5	**20**	**40**
Lecturers: permanent	13	21	34	**62**	**44**
Lecturers: fixed term	0	0	0	**0**	**49**
Total staff	27	24	51	**47**	**39** ·

undergraduate course considering the law from a gender/feminist perspective:
Gender and the Law

undergraduate courses considering the law from a gender/feminist perspective in parts:
Medicine, Law and Ethics
Criminology

postgraduate degree/course/s considering the law from a gender/feminist perspective: **none**

The university has a **Centre for Law and Gender Studies**

Brunel University, Department of Law

	Men	Women	Total	% Women	UK % Women
Professoriat	4	1	5	**20**	**14**
Senior	3	4	7	**57**	**40**
Lecturers: permanent	6	4	10	**40**	**44**
Lecturers: fixed term	0	1	1	**0**	**49**
Total staff	13	10	23	**43**	**39**

undergraduate course considering the law from a gender/feminist perspective:
Law, State and Gender

undergraduate course/s considering the law from a gender/feminist perspective in parts: **none**

postgraduate degree in **Child Law and Policy** *offered*

University of Buckingham, School of Law

	Men	Women	Total	% Women	UK% Women
Professoriat	3	0	3	0	14
Senior	4	3	7	43	40
Lecturers: permanent	0	0	0	0	44
Lecturers: fixed term	5	4	9	44	49
Total staff	12	7	19	37	39

undergraduate course considering the law from a gender/feminist perspective:
Sex and Gender in the Legal Process

*undergraduate courses considering the law from a gender/feminist perspective
in parts:*
Human Rights
Criminology

*postgraduate degree/course/s considering the law from a gender/feminist
perspective:* **none**

University of Cambridge, School of Law

	Men	Women	Total	% Women	UK % Women
Professoriat	16	0	16	0	14
Senior	7	2	9	22	40
Lecturers: permanent	21	8	29	28	44
Lecturers: fixed term	3	2	5	40	49
Total staff	47	12	59	20	39

undergraduate course considering the law from a gender/feminist perspective:
Women and the Law

*undergraduate course/s considering the law from a gender/feminist perspective
in parts:* **none**

*postgraduate degree/course/s considering the law from a gender/feminist
perspective:* **none**

University of Wales, Cardiff Law School

	Men	Women	Total	% Women	UK % Women
Professoriat	9	1	10	**10**	**14**
Senior	9	2	11	**18**	**40**
Lecturers: permanent	30	2	32	**6**	**44**
Lecturers: fixed term	5	0	5	**0**	**49**
Total staff	53	5	58	**9**	**39**

undergraduate course/s considering the law from a gender/feminist perspective: **none**

undergraduate courses considering the law from a gender/feminist perspective in parts:
Advanced Criminal Law
Sociology of Law
Civil Liberties
Legal Philosophies

postgraduate degree/course/s considering the law from a gender/feminist perspective: **none**

University of Central England in Birmingham, Faculty of Law and Social Science

	Men	Women	Total	% Women	UK % Women
Professoriat	0	0	0	**0**	**14**
Senior	16	11	27	**41**	**40**
Lecturers: permanent	0	0	0	**0**	**44**
Lecturers: fixed term	0	0	0	**0**	**49**
Total staff	16	11	27	**41**	**39**

undergraduate course/s considering the law from a gender/feminist perspective: **none**

undergraduate courses considering the law from a gender/feminist perspective in parts:
Criminology
Jurisprudence
Property Law

postgraduate degree/course/s considering the law from a gender/feminist perspective: **none**

University of Central Lancashire, Department of Legal Studies

	Men	Women	Total	% Women	UK % Women
Professoriat	1	0	1	0	14
Senior	12	10	22	45	40
Lecturers: permanent	2	8	10	80	44
Lecturers: fixed term	1	1	2	50	49
Total staff	16	19	35	54	39

undergraduate course/s considering the law from a gender/feminist perspective: **none**

undergraduate courses considering the law from a gender/feminist perspective in parts:
Criminology
Jurisprudence

postgraduate course on LLM in Employment Law considering the law from a gender/feminist perspective in part:
Discrimination Law

City University, Department of Law

	Men	Women	Total	% Women	UK % Women
Professoriat	11	1	12	8	14
Senior	3	0	3	0	40
Lecturers: permanent	2	1	3	33	44
Lecturers: fixed term	7	6	13	46	49
Total staff	23	8	31	26	39

undergraduate course/s considering the law from a gender/feminist perspective: **none**

undergraduate course considering the law from a gender/feminist perspective in part:
Jurisprudence

postgraduate degree/course/s considering the law from a gender/feminist perspective: **none**

University of Coventry, School of International Studies and Law

	Men	Women	Total	% Women	UK % Women
Professoriat	0	0	0	0	14
Senior	8	10	18	56	40
Lecturers: permanent	0	1	1	100	44
Lecturers: fixed term	0	1	1	100	49
Total staff	8	12	20	60	39

undergraduate course considering the law from a gender/feminist perspective: **Women and the Law**

undergraduate course/s considering the law from a gender/feminist perspective in parts: **none**

postgraduate degree/course/s considering the law from a gender/feminist perspective: **none**

A **Women's Studies Centre** *is located in the School of Health and Social Studies*

De Montfort University, School of Law

	Men	Women	Total	% Women	UK % Women
Professoriat	8	0	8	0	14
Senior	12	15	27	56	40
Lecturers: permanent	3	4	7	57	44
Lecturers: fixed term	0	1	1	100	49
Total staff	23	20	43	47	39

undergraduate course/s considering the law from a gender/feminist perspective: **none**

undergraduate course/s considering the law from a gender/feminist perspective in parts: **none**

postgraduate degree/course/s considering the law from a gender/feminist perspective: **none**

University of Derby, Law Division, School of European & International Studies

	Men	Women	Total	% Women	UK % Women
Professoriat	1	0	1	0	14
Senior	4	5	9	56	40
Lecturers: permanent	4	1	5	20	44
Lecturers: fixed term	0	0	0	0	49
Total staff	9	6	15	40	39

undergraduate course/s considering the law from a gender/feminist perspective: none

undergraduate course/s considering the law from a gender/feminist perspective in parts: none

postgraduate degree/course/s considering the law from a gender/feminist perspective: none

University of Dundee, Department of Law

	Men	Women	Total	% Women	UK % Women
Professoriat	6	0	6	0	14
Senior	3	4	7	57	40
Lecturers: permanent	6	3	9	33	44
Lecturers: fixed term	0	1	1	100	49
Total staff	15	8	23	35	39

undergraduate course considering the law from a gender/feminist perspective: Gender and the Law

undergraduate course/s considering the law from a gender/feminist perspective in parts: none

postgraduate degree/course/s considering the law from a gender/feminist perspective: none

University of Durham, Department of Law

	Men	Women	Total	% Women	UK % Women
Professoriat	4	1	5	20	14
Senior	2	1	3	33	40
Lecturers: permanent	3	3	6	50	44
Lecturers: fixed term	2	2	4	50	49
Total staff	11	7	18	39	39

undergraduate course/s considering the law from a gender/feminist perspective: **none**

undergraduate courses considering the law from a gender/feminist perspective in parts:
EC Social Policy Law
Civil Liberties

postgraduate course on LLM considering the law from a gender/feminist perspective in part:
EC Social Policy Law

University of East London, Department of Law

	Men	Women	Total	% Women	UK % Women
Professoriat	1	1	2	50	14
Senior	7	12	19	63	40
Lecturers: permanent	4	2	6	33	44
Lecturers: fixed term	4	1	5	20	49
Total staff	16	16	32	50	39

undergraduate course considering the law from a gender/feminist perspective:
Women and the Law

undergraduate courses considering the law from a gender/feminist perspective in parts:
Legal Cultures
International Human Rights
Legal Theory
Crime and Society
and a large number of the core subjects

postgraduate courses on the LLM considering the law from a gender/feminist perspective in parts:
International Human Rights
Women and the Law
Legal Theory

University of Edinburgh, Faculty of Law

	Men	Women	Total	% Women	UK % Women
Professoriat	13	0	13	**0**	**14**
Senior	10	2	12	**17**	**40**
Lecturers: permanent	8	8	16	**50**	**44**
Lecturers: fixed term	2	1	3	**33**	**49**
Total staff	33	11	44	**25**	**39**

undergraduate course considering the law from a gender/feminist perspective:
Gender and Justice

undergraduate courses considering the law from a gender/feminist perspective in parts:
Criminology
Jurisprudence
Punishment and Society
Human Rights

postgraduate courses considering the law from a gender/feminist perspective in parts:
MSc in Gender, Crime and Criminal Justice
LLM in European Human Rights

Essex University, School of Law

	Men	Women	Total	% Women	UK % Women
Professoriat	6	2	8	**25**	**14**
Senior	13	5	18	**28**	**40**
Lecturers: permanent	0	0	0	**0**	**44**
Lecturers: fixed term	0	0	0	**0**	**49**
Total staff	19	7	26	**27**	**39**

undergraduate course/s considering the law from a gender/feminist perspective: **none**

undergraduate course/s considering the law from a gender/feminist perspective in parts: **none**

postgraduate degree/course/s considering the law from a gender/feminist perspective: **none**

University of Exeter, Faculty of Law

	Men	Women	Total	% Women	UK % Women
Professoriat	2	2	4	**50**	**14**
Senior	9	1	10	**10**	**40**
Lecturers: permanent	3	5	8	**63**	**44**
Lecturers: fixed term	1	3	4	**75**	**49**
Total staff	15	11	26	**42**	**39**

undergraduate course/s considering the law from a gender/feminist perspective: **none**

undergraduate course considering the law from a gender/feminist perspective in part:
Philosophy of Law

postgraduate degree/course/s considering the law from a gender/feminist perspective: **none**

University of Glamorgan, School of Law

	Men	Women	Total	% Women	UK % Women
Professoriat	0	1	1	**100**	**14**
Senior	12	16	28	**57**	**40**
Lecturers: permanent	0	0	0	**0**	**44**
Lecturers: fixed term	4	1	5	**20**	**49**
Total staff	16	18	34	**53**	**39**

undergraduate course/s considering the law from a gender/feminist perspective: **none**

undergraduate course/s considering the law from a gender/feminist perspective in parts: **none**

postgraduate degree/course/s considering the law from a gender/feminist perspective: **none**

Glasgow University, Faculty of Law

	Men	Women	Total	% Women	UK % Women
Professoriat	6	3	9	33	14
Senior	4	5	9	56	40
Lecturers: permanent	6	10	16	63	44
Lecturers: fixed term	0	0	0	0	49
Total staff	16	18	34	53	39

undergraduate course/s considering the law from a gender/feminist perspective: **none**

undergraduate course considering the law from a gender/feminist perspective in part:
Jurisprudence

postgraduate courses on Human Rights on the LLM consider **anti-discrimination** *protections*

Glasgow Caledonian University, Department of Law and Public Administration

	Men	Women	Total	% Women	UK % Women
Professoriat	0	0	0	0	14
Senior	1	2	3	67	40
Lecturers: permanent	8	3	11	27	44
Lecturers: fixed term	2	1	3	33	49
Total staff	11	6	17	35	39

undergraduate course/s considering the law from a gender/feminist perspective: **none**

undergraduate course considering the law from a gender/feminist perspective in part:
Jurisprudence

postgraduate degree/course/s considering the law from a gender/feminist perspective: **none**

University of Greenwich, Law School

	Men	Women	Total	% Women	UK % Women
Professoriat	1	0	1	**0**	**14**
Senior	19	6	25	**24**	**40**
Lecturers: permanent	0	0	0	**0**	**44**
Lecturers: fixed term	1	0	1	**0**	**49**
Total staff	21	6	27	**22**	**39**

undergraduate course/s considering the law from a gender/feminist perspective: **none**

undergraduate course/s considering the law from a gender/feminist perspective in parts: **none**

Establishment of a BA in Criminal Justice Studies, including a feminist perspective, is being considered

postgraduate degree/course/s considering the law from a gender/feminist perspective: **none**

Heriot Watt University, School of Management

	Men	Women	Total	% Women	UK % Women
Professoriat	1	0	1	**0**	**14**
Senior	0	0	0	**0**	**40**
Lecturers: permanent	0	3	3	**100**	**44**
Lecturers: fixed term	0	0	0	**0**	**49**
Total staff	1	3	4	**75**	**39**

undergraduate course/s considering the law from a gender/feminist perspective: **none**

undergraduate course/s considering the law from a gender/feminist perspective in parts: **none**

postgraduate degree/course/s considering the law from a gender/feminist perspective: **none**

University of Hertfordshire, Department of Law

	Men	Women	Total	% Women	UK % Women
Professoriat	0	2	2	100	14
Senior	11	12	23	52	40
Lecturers: permanent	0	3	3	100	44
Lecturers: fixed term	2	3	5	60	49
Total staff	13	20	33	61	39

undergraduate course/s considering the law from a gender/feminist perspective: **none**

undergraduate course considering the law from a gender/feminist perspective in part:
Legal Theory

postgraduate degree considering the law from a gender/feminist perspective:
LLM in Gender and the Law

University of Huddersfield, Law/Business School

	Men	Women	Total	% Women	UK % Women
Professoriat	0	1	1	100	14
Senior	10	7	17	41	40
Lecturers: permanent	0	2	2	100	44
Lecturers: fixed term	0	0	0	0	49
Total staff	10	10	20	50	39

undergraduate course/s considering the law from a gender/feminist perspective: **none**

undergraduate course/s considering the law from a gender/feminist perspective in part: **discrimination law** *is studied as part of the Employment Law course*

postgraduate degree/course/s considering the law from a gender/feminist perspective: **none**

University of Hull, School of Law

	Men	Women	Total	% Women	UK% Women
Professoriat	4	0	4	**0**	**14**
Senior	6	1	7	**14**	**40**
Lecturers: permanent	7	3	10	**30**	**44**
Lecturers: fixed term	2	4	6	**67**	**49**
Total staff	19	8	27	**30**	**39**

undergraduate course/s considering the law from a gender/feminist perspective: **none**

undergraduate course/s considering the law from a gender/feminist perspective in parts: **none**

postgraduate degree/course/s considering the law from a gender/feminist perspective: **none**

The university has a **Hull Centre for Gender Studies**

University of Keele, School of Law

	Men	Women	Total	% Women	UK % Women
Professoriat	2	1	3	**33**	**14**
Senior	3	1	4	**25**	**40**
Lecturers: permanent	4	5	9	**56**	**44**
Lecturers: fixed term	0	0	0	**0**	**49**
Total staff	9	7	16	**44**	**39**

undergraduate course considering the law from a gender/feminist perspective:
Women and the Law

undergraduate course considering the law from a gender/feminist perspective in part:
Jurisprudence

postgraduate degree/course/s considering the law from a gender/feminist perspective: **none**

The university has a **Gender, Sexuality and the Law Group**

University of Kent at Canterbury, Kent Law School

	Men	Women	Total	% Women	UK % Women
Professoriat	3	0	3	0	14
Senior	4	3	7	43	40
Lecturers: permanent	10	8	18	44	44
Lecturers: fixed term	0	0	0	0	49
Total staff	17	11	28	39	39

undergraduate course considering the law from a gender/feminist perspective: **Feminist Perspectives of the Law** (not run for three years)

undergraduate course/s considering the law from a gender/feminist perspective in parts: All courses incorporate feminist perspective as deemed appropriate by course convenors

postgraduate degree/course/s considering the law from a gender/feminist perspective: **none**

Kent Law School is the editorial home of the journal **Feminist Legal Studies**

King's College, London, School of Law

	Men	Women	Total	% Women	UK % Women
Professoriat	8	2	10	20	14
Senior	14	10	24	42	40
Lecturers: permanent	0	0	0	0	44
Lecturers: fixed term	7	5	12	42	49
Total staff	29	17	46	37	39

undergraduate course/s considering the law from a gender/feminist perspective: **none**

undergraduate course/s considering the law from a gender/feminist perspective in part:
Jurisprudence and Legal Theory

postgraduate courses on LLM considering the law from a gender/feminist perspective in parts:
Human Rights of Women
Equality and the Law

Kingston University, Kingston Law School

	Men	Women	Total	% Women	UK % Women
Professoriat	1	0	1	**0**	**14**
Senior	6	10	16	**63**	**40**
Lecturers: permanent	2	0	2	**0**	**44**
Lecturers: fixed term	0	1	1	**100**	**49**
Total staff	9	11	20	**55**	**39**

undergraduate course/s considering the law from a gender/feminist perspective: **none**

undergraduate course/s considering the law from a gender/feminist perspective in parts: **none**

postgraduate degree/course/s considering the law from a gender/feminist perspective: **none**

Lancaster University, Department of Law

	Men	Women	Total	% Women	UK % Women
Professoriat	3	0	3	**0**	**14**
Senior	2	0	2	**0**	**40**
Lecturers: permanent	7	5	12	**42**	**44**
Lecturers: fixed term	0	1	1	**100**	**49**
Total staff	12	6	18	**33**	**39**

undergraduate course considering the law from a gender/feminist perspective:
Gender and the Law

undergraduate courses considering the law from a gender/feminist perspective in parts:
Jurisprudence
Family Law
Criminal Law

postgraduate course on LLM considering the law from a gender/feminist perspective:
Women, Gender and the Legal Order

Leeds University, Department of Law

	Men	Women	Total	% Women	UK % Women
Professoriat	5	2	7	29	14
Senior	3	1	4	25	40
Lecturers: permanent	7	7	14	50	44
Lecturers: fixed term	4	0	4	0	49
Total staff	19	10	29	34	39

undergraduate course/s considering the law from a gender/feminist perspective: **none**

undergraduate course considering the law from a gender/feminist perspective in part: **Jurisprudence**

postgraduate courses considering the law from a gender/feminist perspective: **LLM courses can be taken from the Institute for Gender Studies within the School of Social Sciences and Law**

Leeds Metropolitan University, Law School

	Men	Women	Total	% Women	UK % Women
Professoriat	0	0	0	0	14
Senior	12	13	25	52	40
Lecturers: permanent	0	0	0	0	44
Lecturers: fixed term	0	2	2	100	49
Total staff	12	15	27	56	39

undergraduate course/s considering the law from a gender/feminist perspective: **none**

undergraduate course/s considering the law from a gender/feminist perspective in parts: **none**

postgraduate course on LLM considering the law from a gender/feminist perspective:
Gender and the Law

University of Leicester, Department of Law

	Men	Women	Total	% Women	UK % Women
Professoriat	6	0	6	**0**	**14**
Senior	10	2	12	**17**	**40**
Lecturers: permanent	5	5	10	**50**	**44**
Lecturers: fixed term	10	3	13	**23**	**49**
Total staff	31	10	41	**24**	**39**

undergraduate course/s considering the law from a gender/feminist perspective: **none**

undergraduate course considering the law from a gender/feminist perspective in part:
Criminology

postgraduate course on LLM considering the law from a gender/feminist perspective in part:
Criminology

Host of annual essay-writing competition for the best essay on **Women and the Law**

University of Lincolnshire and Humberside, Department of Law

	Men	Women	Total	% Women	UK % Women
Professoriat	0	0	0	**0**	**14**
Senior	3	3	6	**50**	**40**
Lecturers: permanent	0	0	0	**0**	**44**
Lecturers: fixed term	1	2	3	**67**	**49**
Total staff	4	5	9	**56**	**39**

undergraduate course/s considering the law from a gender/feminist perspective: **none**

undergraduate course/s considering the law from a gender/feminist perspective in parts: **none**

postgraduate degree/course/s considering the law from a gender/feminist perspective: **none**

University of Liverpool, Faculty of Law

	Men	Women	Total	% Women	UK % Women
Professoriat	3	3	6	50	14
Senior	2	3	5	60	40
Lecturers: permanent	2	2	4	50	44
Lecturers: fixed term	5	6	11	55	49
Total staff	12	14	26	54	39

undergraduate course considering the law from a gender/feminist perspective: **Law and the Sexes**

undergraduate course/s considering the law from a gender/feminist perspective in parts: **none**

postgraduate degree/course/s considering the law from a gender/feminist perspective: **none**

The Faculty has a **Feminist Legal Research Unit**

Liverpool John Moores University, School of Law, Social Work and Social Policy

	Men	Women	Total	% Women	UK % Women
Professoriat	1	0	1	0	14
Senior	11	7	18	39	40
Lecturers: permanent	0	3	3	100	44
Lecturers: fixed term	0	0	0	0	49
Total staff	12	10	22	45	39

undergraduate course/s considering the law from a gender/feminist perspective: **none**

undergraduate course/s considering the law from a gender/feminist perspective in parts: **none**

postgraduate degree/course/s considering the law from a gender/feminist perspective: **none**

London Guildhall University, Department of Law

	Men	Women	Total	% Women	UK % Women
Professoriat	1	0	1	0	14
Senior	23	14	37	38	40
Lecturers: permanent	3	3	6	50	44
Lecturers: fixed term	10	9	19	47	49
Total staff	37	26	63	41	39

undergraduate course/s considering the law from a gender/feminist perspective: **none**

undergraduate course considering the law from a gender/feminist perspective in part:
Legal Theory

postgraduate degree/course/s considering the law from a gender/feminist perspective: **none**

London School of Economics, Department of Law

	Men	Women	Total	% Women	UK % Women
Professoriat	8	2	10	20	14
Senior	8	5	13	38	40
Lecturers: permanent	8	3	11	27	44
Lecturers: fixed term	0	0	0	0	49
Total staff	24	10	34	29	39

undergraduate course/s considering the law from a gender/feminist perspective: **none**

undergraduate course/s considering the law from a gender/feminist perspective in parts: **none**

postgraduate course on LLM considering the law from a gender/feminist perspective in part:
Human Rights of Women

The university has a **Gender Institute**

University of Manchester, Faculty of Law

	Men	Women	Total	% Women	UK % Women
Professoriat	7	1	8	13	14
Senior	7	2	9	22	40
Lecturers: permanent	4	9	13	69	44
Lecturers: fixed term	1	1	2	50	49
Total staff	19	13	32	41	39

undergraduate course considering the law from a gender/feminist perspective:
Gender and the Law

undergraduate course/s considering the law from a gender/feminist perspective in parts: **none**

postgraduate courses considering the law from a gender/feminist perspective:
Gender and Sexuality in the Criminal Justice System *is offered as part of the* **MA in Crime, Law and Society; Gender, Medicine and the Law** *is offered as part of the* **MA in Health Care Ethics**

Course in Employment Equality Law is being planned for the taught LLM

Manchester Metropolitan University, School of Law

	Men	Women	Total	% Women	UK % Women
Professoriat	1	0	1	0	14
Senior	27	21	48	44	40
Lecturers: permanent	3	2	5	40	44
Lecturers: fixed term	0	0	0	0	49
Total staff	31	23	54	43	39

undergraduate course considering the law from a gender/feminist perspective:
Gender and the Law

undergraduate courses considering the law from a gender/feminist perspective in parts:
Criminology
Law in Contemporary Society

postgraduate courses considering the law from a gender/feminist perspective in parts: **all MA courses include equal opportunities issues, eg: in sports law, employment law, criminal law, EU law**

Middlesex University, Law/Business School

	Men	Women	Total	% Women	UK % Women
Professoriat	1	1	2	**50**	**14**
Senior	7	8	15	**53**	**40**
Lecturers: permanent	0	4	4	**100**	**44**
Lecturers: fixed term	0	0	0	**0**	**49**
Total staff	8	13	21	**62**	**39**

undergraduate course considering the law from a gender/feminist perspective:
Women and the Law

undergraduate courses considering the law from a gender/feminist perspective in parts:
Criminal Justice and Procedure
Law and Society
Criminal Law

postgraduate degree/course/s considering the law from a gender/feminist perspective: **none**

Napier University, Department of Law

	Men	Women	Total	% Women	UK % Women
Professoriat	0	0	0	**0**	**14**
Senior	2	2	4	**50**	**40**
Lecturers: permanent	9	6	15	**40**	**44**
Lecturers: fixed term	0	0	0	**0**	**49**
Total staff	11	8	19	**42**	**39**

undergraduate course/s considering the law from a gender/feminist perspective: **none**

undergraduate course/s considering the law from a gender/feminist perspective in parts: **none**

postgraduate course on LLM in International Law considering the law from a gender/feminist perspective in part:
Human Rights Law

University of Newcastle upon Tyne, Newcastle Law School

	Men	Women	Total	% Women	UK % Women
Professoriat	3	0	3	0	14
Senior	5	1	6	17	40
Lecturers: permanent	6	4	10	40	44
Lecturers: fixed term	0	1	1	100	49
Total staff	14	6	20	30	39

undergraduate course/s considering the law from a gender/feminist perspective: **none**

undergraduate course considering the law from a gender/feminist perspective in parts:
Criminology
Current Issues in Jurisprudence

postgraduate degree/course/s considering the law from a gender/feminist perspective: **none**

The university has a **Centre for Gender and Women's Studies**

University of North London, School of Law

	Men	Women	Total	% Women	UK % Women
Professoriat	0	0	0	0	14
Senior	7	6	13	46	40
Lecturers: permanent	2	2	4	50	44
Lecturers: fixed term	0	0	0	0	49
Total staff	9	8	17	47	39

undergraduate course/s considering the law from a gender/feminist perspective: **none**

undergraduate course considering the law from a gender/feminist perspective in part:
Jurisprudence

postgraduate degree/course/s considering the law from a gender/feminist perspective: **none**

University of Northumbria at Newcastle, School of Law

	Men	Women	Total	% Women	UK % Women
Professoriat	1	0	1	**0**	**14**
Senior	55	26	81	**32**	**40**
Lecturers: permanent	0	0	0	**0**	**44**
Lecturers: fixed term	0	0	0	**0**	**49**
Total staff	56	26	82	**32**	**39**

undergraduate course/s considering the law from a gender/feminist perspective: **none**

undergraduate courses considering the law from a gender/feminist perspective in parts:
Crime, Law and Society
Jurisprudence

postgraduate degree/course/s considering the law from a gender/feminist perspective: **none**

University of Nottingham, Department of Law

	Men	Women	Total	% Women	UK % Women
Professoriat	7	1	8	**13**	**14**
Senior	6	1	7	**14**	**40**
Lecturers: permanent	10	7	17	**41**	**44**
Lecturers: fixed term	0	0	0	**0**	**49**
Total staff	23	9	32	**28**	**39**

undergraduate course/s considering the law from a gender/feminist perspective: **none**

undergraduate courses considering the law from a gender/feminist perspective in parts:
Criminology
Mental Health Law
Health Care Law
Understanding Law

postgraduate degree/course/s considering the law from a gender/feminist perspective: **none**

Nottingham Trent University, Nottingham Law School

	Men	Women	Total	% Women	UK % Women
Professoriat	1	0	1	0	14
Senior	25	16	41	39	40
Lecturers: permanent	0	0	0	0	44
Lecturers: fixed term	3	5	8	63	49
Total staff	29	21	50	42	39

undergraduate course considering the law from a gender/feminist perspective:
Sexuality and the Law

*undergraduate course/s considering the law from a gender/feminist perspective
in parts:* **none**

*postgraduate degree/course/s considering the law from a gender/feminist
perspective:* **none**

Oxford Brookes University, School of Social Sciences and Law

	Men	Women	Total	% Women	UK % Women
Professoriat	0	0	0	0	14
Senior	5	9	14	64	40
Lecturers: permanent	0	0	0	0	44
Lecturers: fixed term	0	0	0	0	49
Total staff	5	9	14	64	39

undergraduate course considering the law from a gender/feminist perspective:
Sexuality and the Law

*undergraduate courses considering the law from a gender/feminist perspective
in parts:*
Discrimination Law
Crime and Society
Law, Rights and the Individual

*postgraduate degree/course/s considering the law from a gender/feminist
perspective:* **none**

University of Plymouth, Plymouth Business School

	Men	Women	Total	% Women	UK % Women
Professoriat	1	0	1	0	14
Senior	7	6	13	46	40
Lecturers: permanent	0	1	1	100	44
Lecturers: fixed term	0	2	2	100	49
Total staff	8	9	17	53	39

undergraduate course/s considering the law from a gender/feminist perspective: **none**

undergraduate course/s considering the law from a gender/feminist perspective in parts: **none**

postgraduate degree/course/s considering the law from a gender/feminist perspective: **none**

Queen's University, Belfast, School of Law

	Men	Women	Total	% Women	UK % Women
Professoriat	4	2	6	33	14
Senior	11	2	13	15	40
Lecturers: permanent	7	9	16	56	44
Lecturers: fixed term	0	0	0	0	49
Total staff	22	13	35	37	39

undergraduate course/s considering the law from a gender/feminist perspective:
Gender and the Law *(not run since 1995)*

undergraduate course/s considering the law from a gender/feminist perspective in parts: **none**

postgraduate courses on **LLM in Human Rights** *consider the law from a gender/ feminist perspective in parts*

The Centre for Human Rights considers feminist perspectives

University of Reading, Department of Law

	Men	Women	Total	% Women	UK % Women
Professoriat	3	0	3	0	14
Senior	4	3	7	43	40
Lecturers: permanent	4	3	7	43	44
Lecturers: fixed term	0	0	0	0	49
Total staff	11	6	17	35	39

undergraduate course/s considering the law from a gender/feminist perspective: **none**

undergraduate courses considering the law from a gender/feminist perspective in parts:
Jurisprudence
Human Rights

postgraduate degree/course/s considering the law from a gender/feminist perspective: **none**

Robert Gordon University, Department of Public Administration & Law

	Men	Women	Total	% Women	UK % Women
Professoriat	0	0	0	0	14
Senior	3	0	3	0	40
Lecturers: permanent	3	5	8	63	44
Lecturers: fixed term	1	2	3	67	49
Total staff	7	7	14	50	39

undergraduate course/s considering the law from a gender/feminist perspective: **none**

undergraduate course/s considering the law from a gender/feminist perspective in parts: **none**

postgraduate degree/course/s considering the law from a gender/feminist perspective: **none**

School of Oriental & African Studies, University of London, Department of Law

	Men	Women	Total	% Women	UK % Women
Professoriat	2	0	2	0	14
Senior	8	0	8	0	40
Lecturers: permanent	2	0	2	0	44
Lecturers: fixed term	3	5	8	63	49
Total staff	15	5	20	25	39

undergraduate course/s considering the law from a gender/feminist perspective: **none**

undergraduate course/s considering the law from a gender/feminist perspective in parts: **none**

postgraduate course on LLM considering the law from a gender/feminist perspective:
Human Rights of Women

University of Sheffield, Faculty of Law

	Men	Women	Total	% Women	UK % Women
Professoriat	10	2	12	17	14
Senior	4	3	7	43	40
Lecturers: permanent	8	17	25	68	44
Lecturer: fixed term	1	0	1	0	49
Total staff	23	22	45	49	39

undergraduate course considering the law from a gender/feminist perspective:
Women and Crime

undergraduate course/s considering the law from a gender/feminist perspective in parts: **none**

postgraduate degree/course/s considering the law from a gender/feminist perspective: **none**

Sheffield Hallam University, Department of Financial Studies & Law

	Men	Women	Total	% Women	UK % Women
Professoriat	1	0	1	0	14
Senior	9	7	16	44	40
Lecturers: permanent	0	2	2	100	44
Lecturers: fixed term	0	0	0	0	49
Total staff	10	9	19	47	39

undergraduate course/s considering the law from a gender/feminist perspective: **none**

undergraduate course/s considering the law from a gender/feminist perspective in parts: **none**

postgraduate degree/course/s considering the law from a gender/feminist perspective: **none**

University of Southampton, Faculty of Law

	Men	Women	Total	% Women	UK % Women
Professoriat	6	0	6	0	14
Senior	7	2	9	22	40
Lecturers: permanent	6	6	12	50	44
Lecturers: fixed term	0	1	1	100	49
Total staff	19	9	28	32	39

undergraduate course/s considering the law from a gender/feminist perspective: **none**

undergraduate courses considering the law from a gender/feminist perspective in parts:
Law and Discrimination
Criminology

postgraduate degree/course/s considering the law from a gender/feminist perspective: **none**

South Bank University, Department of Legal, Political and Social Sciences

	Men	Women	Total	% Women	UK % Women
Professoriat	0	0	0	0	14
Senior	8	0	8	0	40
Lecturers: permanent	4	5	9	56	44
Lecturers: fixed term	0	3	3	100	49
Total staff	12	8	20	40	39

undergraduate courses considering the law from a gender/feminist perspective:
two Women and the Law units

undergraduate course/s considering the law from a gender/feminist perspective in parts: **none**

postgraduate degree/course/s considering the law from a gender/feminist perspective: **none**

Staffordshire University, School of Law

	Men	Women	Total	% Women	UK % Women
Professoriat	2	0	2	0	14
Senior	15	10	25	40	40
Lecturers: permanent	4	4	8	50	44
Lecturers: fixed term	0	0	0	0	49
Total staff	21	14	35	40	39

undergraduate course considering the law from a gender/feminist perspective:
Feminism, Post Structuralism and the Law

undergraduate course/s considering the law from a gender/feminist perspective in parts: **none**

postgraduate course on LLM considering the law from a gender/feminist perspective:
Feminism, Post Structuralism and the Law

University of Strathclyde, Law School

	Men	Women	Total	% Women	UK % Women
Professoriat	10	0	10	0	14
Senior	3	1	4	25	40
Lecturers: permanent	4	4	8	50	44
Lecturers: fixed term	2	1	3	33	49
Total staff	19	6	25	24	39

undergraduate course/s considering the law from a gender/feminist perspective: **none**

undergraduate courses considering the law from a gender/feminist perspective in parts:
Law and Society
Legal Theory
Criminology

postgraduate degree/course/s considering the law from a gender/feminist perspective: **none**

University of Surrey, Department of International and Linguistics Studies, Law Section

	Men	Women	Total	% Women	UK % Women
Professoriat	0	0	0	0	14
Senior	0	1	1	100	40
Lecturers: permanent	1	2	3	67	44
Lecturers: fixed term	4	3	7	43	49
Total staff	5	6	11	55	39

undergraduate course/s considering the law from a gender/feminist perspective: **none**

undergraduate course/s considering the law from a gender/feminist perspective in parts: **none**

postgraduate degree/course/s considering the law from a gender/feminist perspective: **none**

University of Sussex, Centre for Legal Studies

	Men	Women	Total	% Women	UK % Women
Professoriat	1	1	2	50	14
Senior	2	2	4	50	40
Lecturers: permanent	4	5	9	56	44
Lecturers: fixed term	0	2	2	100	49
Total staff	7	10	17	59	39

undergraduate course/s considering the law from a gender/feminist perspective: **none**

undergraduate course/s considering the law from a gender/feminist perspective in parts: **none**

postgraduate degree/course/s considering the law from a gender/feminist perspective: **none**

University of Teeside, School of Law, Humanities & International Studies

	Men	Women	Total	% Women	UK % Women
Professoriat	0	0	0	0	14
Senior	8	2	10	20	40
Lecturers: permanent	0	3	3	100	44
Lecturers: fixed term	0	0	0	0	49
Total staff	8	5	13	38	39

undergraduate course/s considering the law from a gender/feminist perspective: **none**

undergraduate course/s considering the law from a gender/feminist perspective in parts: **none**

postgraduate degree/course/s considering the law from a gender/feminist perspective: **none**

Thames Valley University, School of Law

	Men	Women	Total	% Women	UK % Women
Professoriat	1	0	1	0	14
Senior	13	22	35	63	40
Lecturers: permanent	0	0	0	0	44
Lecturers: fixed term	0	0	0	0	49
Total staff	14	22	36	61	39

undergraduate course/s considering the law from a gender/feminist perspective: none

undergraduate course/s considering the law from a gender/feminist perspective in parts: none

postgraduate degree/course/s considering the law from a gender/feminist perspective: none

University of Ulster, School of Public Policy, Economics and Law

	Men	Women	Total	% Women	UK % Women
Professoriat	2	0	2	0	14
Senior	3	1	4	25	40
Lecturers: permanent	6	4	10	40	44
Lecturers: fixed term	0	0	0	0	49
Total staff	11	5	16	31	39

undergraduate course/s considering the law from a gender/feminist perspective: none

undergraduate course/s considering the law from a gender/feminist perspective in parts: none

postgraduate degree/course/s considering the law from a gender/feminist perspective: none

University College London, Faculty of Laws

	Men	Women	Total	% Women	UK % Women
Professoriat	15	2	17	12	14
Senior	8	5	13	38	40
Lecturers: permanent	7	3	10	30	44
Lecturers: fixed term	1	0	1	0	49
Total staff	31	10	41	24	39

undergraduate course/s considering the law from a gender/feminist perspective: none

undergraduate course/s considering the law from a gender/feminist perspective in parts: none

postgraduate degree/course/s considering the law from a gender/feminist perspective: none

University of Warwick, School of Law

	Men	Women	Total	% Women	UK % Women
Professoriat	6	1	7	14	14
Senior	7	3	10	30	40
Lecturers: permanent	9	5	14	36	44
Lecturers: fixed term	2	2	4	50	49
Total staff	24	11	35	31	39

undergraduate course considering the law from a gender/feminist perspective:
Women and the Law

undergraduate courses considering the law from a gender/feminist perspective in parts:
Sociology of Law
Legal Theory
Social Theory of Law

postgraduate courses on LLM in Law and Development considering the law from a gender/feminist perspective:
Comparative Perspectives on Gender, Law and Development

The university has a **Centre for the Study of Women and Gender**

University of Westminster, School of Law

	Men	Women	Total	% Women	UK % Women
Professoriat	1	0	1	0	14
Senior	17	14	31	45	40
Lecturers: permanent	0	0	0	0	44
Lecturers: fixed term	0	0	0	0	49
Total staff	18	14	32	44	39

undergraduate course considering the law from a gender/feminist perspective: **Women and the Law**

undergraduate course/s considering the law from a gender/feminist perspective in parts: **none**

postgraduate degree considering the law from a gender/feminist perspective: **LLM in Women and the Law**

University of the West of England, Bristol, Faculty of Law

	Men	Women	Total	% Women	UK % Women
Professoriat	2	0	2	0	14
Senior	8	3	11	27	40
Lecturers: permanent	27	17	44	39	44
Lecturers: fixed term	0	4	4	100	49
Total staff	37	24	61	39	39

undergraduate course considering the law from a gender/feminist perspective: **Gender and the Law**

undergraduate course considering the law from a gender/feminist perspective in part: **Criminology**

postgraduate degree/course/s considering the law from a gender/feminist perspective: **none**

The **Centre for Criminal Justice** *focuses on gender/feminist perspectives*

University of Wolverhampton, School of Legal Studies

	Men	Women	Total	% **Women**	UK % **Women**
Professoriat	1	0	1	**0**	**14**
Senior	18	14	32	**44**	**40**
Lecturers: permanent	3	5	8	**63**	**44**
Lecturers: fixed term	0	0	0	**0**	**49**
Total staff	22	19	41	**46**	**39**

undergraduate course considering the law from a gender/feminist perspective:
Women and the Law

undergraduate courses considering the law from a gender/feminist perspective in parts:
Employment Law
EU Law
Law and Morality
Individual Rights and Personal Freedoms
Crime, Victims and Offenders

postgraduate degree **LLM in International Human Rights** *considers the law from a gender/feminist perspective in parts*

Methodology

The survey was designed and piloted in July 1997 and took the form of a written questionnaire. One copy of the questionnaire, together with a covering letter explaining the purpose of the survey, was sent to named Heads of Law Departments in August 1997, with the return of questionnaires requested by 30 September 1997. In relation to those universities which had not responded, reminder letters, together with another copy of the questionnaire, were sent to named Heads of Law Departments in October 1997, with a third reminder sent in November 1997. The questionnaire was sent to 81 universities offering law degrees or law courses as part of non-law degrees. 75 of those universities responded, and the data from the completed questionnaires is provided in this appendix, in appendix two and is considered in chapters one and two. Six universities did not respond, namely: Anglia Polytechnic University, Aston University, University of East Anglia, Luton University, Oxford University and Queen Mary & Westfield College. Accordingly, the response rate for the survey was 93 per cent. This is a very high response rate for a postal survey of this kind and enables reliable conclusions to be drawn. In addition, it is suggested that the absence of the six universities named above does not materially affect the resulting data. Three quarters (76 per cent) of the questionnaires were completed by Heads of Departments/Deans of Faculties, a further 18 per cent by academic law staff and the remainder by law school administrative staff.

The questionnaire requested information as of 1 October 1997. No attempt has been made to verify independently the information which each law school provided; nor have any amendments been made to the data to take into account any changes since the questionnaire's census date.

Number of women and men academic staff in UK law schools

The first part of the questionnaire sought information on the number of women and men academic staff in each law school. Where the academic law staff are part of a larger department, for example a Department of Law and Finance, information was sought regarding law staff only. Data was requested on numbers of professors, readers, senior lecturers, principal lecturers, and lecturers on fixed term and permanent contracts. Information was also requested on numbers of honorary and visiting professors. The data relating to honorary and visiting professors is not included in the results given above, on the grounds that the numbers may distort the data. In view of the fact that there were few part-time appointments, the figures include full and part-time appointments. The category of 'senior' staff includes readers, senior lecturers and principal lecturers. These separate categories were collapsed for ease of reference, the main objective being to reveal the numbers of non-professorial promoted staff. However, it should be recognised that the designations of seniority vary between universities, and the same title, eg senior lecturer, may connote a different level of seniority between universities. The data includes all staff in the law school and therefore includes those staff who teach on professional courses such as the Legal Practice Course and the Bar Vocational Course. Separate data was not collected in respect of those who taught on such courses as many law teachers are engaged on both undergraduate and professional courses. Some universities volunteered data on numbers of research assistants, but this information is not included as it had not been requested of all universities.

Undergraduate and postgraduate courses and parts of courses considering the law from a gender/feminist perspective in UK law schools

The second part of the questionnaire sought information regarding the law school curriculum, and in particular the extent to which law schools offer any courses considering the law from a gender/feminist perspective either in whole or in part. Information was sought in four categories:

1.   whether an undergraduate module/course which closely corresponds to any of the following descriptions, Women and the Law, Gender and the Law, Feminist Perspectives of Law, Feminist Jurisprudence, is offered to law students;

2.   whether any other modules/courses include, as a specific part of the
      course, a consideration of the law from a gender/feminist perspective
      in part, eg feminist criminology as part of a Criminology module, or
      feminist jurisprudence as part of a Jurisprudence course;
3.   whether a taught postgraduate degree which closely corresponds to
      any of the following, Women and the Law, Feminist Perspectives on
      the Law, Gender and the Law, Feminist Jurisprudence, is offered;
4.   whether any courses/options are available as part of a taught
      postgraduate degree which consider the law from a gender/feminist
      perspective, eg Human Rights of Women as an option on a general
      LLM.

In respect of each category, a number of universities responded that they
were considering introducing courses, and this information is noted in
respect of the relevant university. However, this information is not included
in the calculation of the overall number of that type of course offered, in view
of the fact that the proposal may not come to fruition. Further, as the
questionnaire did not specifically seek information as to whether or not a
course was being offered, some universities may indeed be considering
doing so, but did not include this data on the questionnaire. As regards
category one, the courses offered are named above in respect of each
university and are listed in appendix two. Two universities volunteered
information that the course offered had not run for a few years, and this
information is included above. It may be that other universities do not run
the course listed from year to year because of staff changes, absences or
some other factor. Accordingly, it is not represented that each of the
courses named above runs every year. The total figure of courses offered,
falling within category one, includes the two courses which are offered but
which have not run for a few years.

As regards responses to information falling within category two, no
attempt has been made to verify the extent to which courses consider the
law from a gender/feminist perspective, nor of the value attached to those
parts of the course. The total number of courses which consider the law from
a gender/feminist perspective in part does not include the university which
responded that this perspective is included as deemed appropriate by
course leaders, as this is not quantifiable.

In view of the fact that there are only two universities which offer a degree
falling within category three, the responses for categories three and four are
amalgamated for the purposes of the display of the data above. The
information provided replicates that in the completed questionnaires, and
thus, as with the data in category two, no attempt has been made to verify
the extent to which courses consider the law from a gender/feminist
perspective, nor of the value attached to those parts of the course.

The questionnaire asked whether a law centre/research group/institute
was established which considered the law from a gender/feminist

perspective, information which is included above. Some universities volunteered information that the university hosted an inter-disciplinary centre for gender or women's studies and this information is also included. It may be that many other universities have also established such centres, to which reference was not made on the completed questionnaire, thus the information above should not be read as suggesting that only those universities host such a centre.

Appendix 4
Types of postgraduate law degrees

POSTGRADUATE DEGREES BY RESEARCH

There are a number of postgraduate degrees available which are by research, meaning that assessment will be by submission of a thesis of a specified word length. Such a degree involves a greater requirement of self-direction than a taught postgraduate degree, although each student has a supervisor who will guide the research, particularly in the early stages. Generally, in order to gain admission to a postgraduate degree by research at least a 2.1 degree at undergraduate level will be required. For some degrees, such as a doctorate, a taught postgraduate degree will be an advantage. In particular, as regards funding, a first class degree will be all but a prerequisite. There are a variety of degrees by research which are considered below. Up to date information should be sought from each law school.

Doctorates (PhD/DPhil)

A doctorate is the highest level of postgraduate degree (see further Estelle Phillips and Derek Pugh, *How to get a PhD*). It will generally involve independent research culminating in the writing of a thesis of around 100,000 words. A doctoral thesis must significantly develop research in the chosen field, and will usually require three to four years of full-time study, or five to six year's part-time. Assessment will be made on the basis of the written thesis, with an oral examination, known as a viva, being commonplace. The title of PhD is the usual title for a doctoral degree, but a few institutions offer the degree of DPhil which is equivalent.

MPhil (MLitt)

The MPhil is similar to a doctorate but the thesis will probably only be around 60,000 words. Accordingly, this degree is completed over a shorter time scale, usually two-three year's full-time, or four-five year's part-time. Again, assessment will be by production of a written thesis, with a possibility of a vivar. Many students register initially for an MPhil and then transfer to a PhD after their first year.

LLM (MJur) by research

It is also possible to complete a higher degree by research within one year full-time, or two to three year's part-time. The written thesis required will be proportionately smaller, perhaps around 30,000 words, and a viva is less likely. The principal difference between this degree, and the MPhil and the doctorate, is in the depth of originality of research required. A master's level degree requires a thorough analysis and clear exposition of existing research in the chosen field, whereas a doctoral thesis requires a greater degree of originality constituting a contribution to that research debate.

HIGHER DEGREES BY INSTRUCTION

LLM

A taught LLM degree usually comprises a number of advanced courses combined with some element of coursework. The balance between courses/exams and the dissertation will vary depending on the particular course and institution. An LLM may act as a bridge for a student who wishes to begin a PhD later. Also, it may be that if a student is not sure whether or not they wish to consider further research, or an academic career, the LLM can prove a useful half-way house, as the LLM degree will also be very useful in the legal profession. There are a wide variety of LLM degrees available. Some degrees are specialist in nature, eg European Law or Commercial Law. Both the universities of Westminster and Huddersfield offer an LLM in Women and the Law. Other degrees will be more general in nature and students can choose from a wide variety of options.

Research assistantships

It might be possible, as an alternative to studying for a higher degree, to become a research assistant within a law school, or some other similar institution such as the Law Commission or the Institute of Advanced Legal Studies. Research assistantships may be full time and salaried, and may involve being part of a team carrying out a specific research project. Such a position will enable research experience to be gained, although it may still be possible to register for a higher degree part-time. Another possibility is a research studentship which might entail the law school paying the fees for a higher degree, and perhaps a stipend to the student, with the student undertaking some teaching obligations. Many such jobs are advertised in the national press but may also be sought by approaching the institution or organisation on spec.

Appendix 5
Addresses and contact details

African, Caribbean and Asian Lawyers Group
The Law Society
113 Chancery Lane
London WC2A 1PL
tel: 0171 242 1222
fax: 0171 320 5863

Association of University Teachers
United House
9 Pembridge Road
London W11 3JY
tel: 0171 221 4370
fax: 0171 727 6547
email: hq@aut.org.uk
http://www.aut.org.uk/index.html

Association of Women Barristers
12 Gray's Inn Square
London WC1R 5JP

Association of Women Solicitors
c/o Judith MacDermott
The Law Society
50-52 Chancery Lane
London
WC2A 1SX
tel: 0171 242 1222
maternity helpline: 0181 676 9887
http://www.badgerap.demon.
co.uk/aws/

Bar Lesbian and Gay Group
c/o Chambers of Robert Latham
1 Pump Court
Temple
London EC4Y 7AB
tel: 0171 353 4341

Bar Pro Bono Unit
7 Gray's Inn Square
Gray's Inn
London WC1R 5AZ
tel: 0171 831 9711
fax: 0171 831 9733
DX 188 London Chancery Lane

Bar Equal Opportunity Officers
Bar Sex Discrimination Committee
The Disability Panel
Bar Race Relations Committee
General Council of the Bar
3 Bedford Row
London WC1R 4DB
tel: 0171 242 0082
fax: 0171 831 9217

Education and Training Department
General Council of the Bar
2-3 Cursitor Street
London EC4A 1NE
tel: 0171 440 4000
http://www.barcouncil.org.uk/

CPE/Diploma in Law Central Applications Board
PO Box 84
Guildford
Surrey
GU3 1YX
tel: 01483 451080

Commission for Racial Equality
Headquarters Office
Elliot House
10-12 Allington Street
London SW1E 5EH
tel: 0171 828 7022
http://open.gov.uk/cre/crehome.htm

Committee of Vice Chancellors and Principals
Woburn House
20 Tavistock Square
London WC1H 9HQ
tel: 0171 419 4111
fax: 0171 388 8649
email: info@cvcp.ac.uk
http://www.cvcp.ac.uk

Commission on University Career Opportunity
29 Tavistock Square
London WC1H 9EZ
tel: 0171 419 4111
fax: 0171 388 6256
email: fiona.waye@cvcp.ac.uk
http://www.cvcp.ac.uk/cuco.html

Equal Opportunities Commission
Overseas House
Quay Street
Manchester M3 3HN
tel: 0161 833 9244
http://www.eoc.org.uk/index.htm

Equal Opportunities Pro Bono Group
The Law Society, tel: 0171 320 5800
The Bar Council, tel: 0171 242 0082.

Freelance Solicitors Group
(write enclosing a stamped SAE)
N M Siabkin
5, The Link
West Acton
London W3 0JW

Free Representation Unit (Bar)
tel: 0171 831 0692

Gray's Inn
Education Department
8 South Square
Gray's Inn
London WC1R 5EU
tel: 0171 405 8164

Inner Temple
Education and Training Department
Treasurer's Office
Inner Temple
London EC4Y 7HL
tel: 0171 797 8250

Institute of Legal Executives
Kempston Manor
Kempston
Bedford MK42 7AB
tel: 01234 841 000
DX: 124780 Kempston 2

Law Society's Careers and Recruitment Service
227-228 Strand
London WC2R 1BA
tel: 0171 242 1222
http://www.lawsociety.org.uk

The Law Society
Jerry Garvey
Ethnic Minorities Careers Officer
50-52 Chancery Lane
London
WC2A 1SX
tel: 0171 242 1222

Judicial Appointments Division
Lord Chancellor's Department
Selbourne House
54-60 Victoria Street
London SW1E 6QW
tel: 0171 210 8923
http://www.open.gov.uk/lcd/
lcdhome.htm

Legal Practice Course
Central Applications Board
PO Box 84
Guildford
Surrey
GU3 1YX
tel: 01483 301282

Lincoln's Inn
Students' Department
Treasury Office
Lincoln's Inn
London WC2A 3TL
tel: 0171 405 0138

Maternity Alliance
(advice and information on maternity
and pregnancy rights)
45 Beech Street
London EC2P 2LX
tel (office): 0171 588 8583
fax: 0171 588 8584
information line: 0171 588 8582
e-mail: ma@mail.pro-net.co.uk

Middle Temple
Students' Department
Treasury Office
Middle Temple
London EC4Y 9AT
tel: 0171 427 4800

**National Association of Teachers
and Lecturers in Further and
Higher Education**
27 Britannia Street
London WC1X 9JP
tel: 0171 837 4403

New Ways to Work
(advice on all aspects of flexible working
arrangements)
309 Upper Street
London N1 2TY
tel: 0171 226 4026

National Union of Students
Nelson Mandela House
461 Holloway Road
London N7 6LJ
tel: 0171 272 8900
fax: 0171 263 5713
email: nusuk@nus.org.uk
http://www.nus.org.uk

**Office for the Supervision
of Solicitors**
Victoria Court
8 Dormer Place
Leamington Spa
Warwickshire CV32 5AE
tel: 01926 820 082/3
fax: 01926 431 435

Opportunity 2000
Business in the Community
44 Baker Street
London W1M 1DH
tel: 0171 224 1600
fax: 0171 486 1700

Rights of Women
52-54 Featherstone Street
London EC1Y 8RT
tel: 0171 251 6575
fax: 0171 608 0928
advice line: 0171 251 6577

Society of Black Lawyers
Room 9
Winchester House
11 Cranmer Road
Kennington Park
London SW9 6EJ
tel: 0171 735 6592 (admin)
tel: 0171 735 6591 (advice)
fax: 0171 735 6594

Solicitors' Pro Bono Group
15 St Swithin's Lane
London EC4
tel: 0171 929 6601/2/3.

UCAS
Fulton House
Jessop Avenue
Cheltenham
Gloucestershire GL50 3SH
tel 01242 22 77 88
minicom for hard of hearing: 01242 225857
e-mail: R+S@ucas.ac.uk
http://www.ucas.ac.uk

Young Women Lawyers
c/o Samantha Steer
CCH Editions Ltd
Telford Road
Bicester
Oxfordshire OX6 0XD
tel: 01869 253300
fax: 01869 874702
email: steers@cch.co.uk

Appendix 6
Professional groups and networks

This appendix outlines the role and activities of a number of professional groups in which women lawyers may be interested. Contact details for each of the groups can be found in appendix five. This list is not exhaustive as there are many associations which will be of interest, from other law-related groups focusing on an area of law, or type of practice, as well as cross-professional groups.

Professional support groups and networks play a number of important roles. The networking involved in becoming a member can help an individual gain confidence through the opportunity to share experiences or discuss problems. Especially in a male-dominated profession like law, where many women may feel isolated in firms or chambers, the opportunity to meet with other lawyers in a supportive environment is invaluable. Support through mentoring, both formal and informal, can help career development. The group can act as a pool of contacts or as a source of information and is an accepted place in which to seek advice without embarrassment. Membership and involvement in such groups is not to everyone's taste but can be very rewarding.

The 1992 *Without Prejudice?* research (see chapters five and six) on the Bar and judiciary noted that although the legal profession is extremely competitive, 'it is important that women continue to develop information and support networks'. The report continued that networking takes time and that some women may not see it as a priority, but it emphasised that the provision of self-managed support systems could reduce any feelings of isolation and eventually may influence the culture of legal practice. Finally, it argued that such networks can help women both remain in practice and advance their careers.

However, the groups discussed below do not just serve their members, but, to varying degrees, are active in campaigning to improve the position and status of women in the profession and the law in general. Many of the chapters of this book have discussed the research and memoranda published by these groups which have stimulated and directed debate within the legal profession. These groups, therefore, play a vital role in ensuring that the legal profession as a whole is more open and accountable which better serves the interests of its clients and public. As Fiona Boyle, a former chair of the Trainee Solicitors Group, argues, such groups offer a 'tremendous potential for pressure to be brought to bear on the profession to address the underlying reasons for continuing inequalities'.

African, Caribbean and Asian Lawyers Group

The ACA Lawyers Group was set up in 1991 and aims to provide practical assistance and encouragement to students on all aspects of entrance into the profession, as well as working to raise the profile of minority lawyers within the profession as a whole. It organises lectures, conferences and social events, as well as helping with gaining vacation placements and mini-pupillages. It also runs a mentoring scheme and publishes a magazine for members three times a year. Membership is open to all and is not limited to those of African, Caribbean and Asian descent.

Association of Women Barristers

The AWB represents and promotes the interests of women barristers. Formed in 1991, it organises social and educational events which put women barristers in touch with each other, and with other lawyers, as well as campaigning on behalf of women barristers for equal treatment within the profession. Membership of the AWB is open to all women barristers, whether in private practice or employment, to pupils, squatters and Bar students. The President is Mrs Justice Arden and the present chairwoman is Josephine Hayes.

Association of Women Solicitors

The AWS was formed in 1919 when women were first permitted to qualify as solicitors and represents the views of women solicitors, promotes women's interests in the profession, and campaigns on issues relevant to women solicitors. The AWS sees itself partly as a club where women can mix informally, in a business environment. It is recognised by, and receives funding from, the Law Society. The group holds a number of conferences, social events and residential weekends and arranges a number of courses, such as the annual course for women intending to return to practice having taken a career break. The group has also produced many documents and memoranda to help women in their careers such as draft clauses for partnership deeds covering maternity and advice for women wishing to work part-time. In order to offer some positive role models and advice to women, the AWS established a mentoring scheme in 1996. Under the scheme, more experienced solicitors will make themselves available to advise and support their junior colleagues across a range of issues. Any woman solicitor, trainee or woman who has passed the LPC and is seeking a training contract may join the group, and there is no fee. There are a wide number of regional groups of the AWS which organise local events. The present chairwoman is Judith Willis who has also run the group's Maternity Helpline since it was established in 1994 to help women solicitors whose careers were threatened by the attitudes of some firms towards maternity.

Rights of Women

Whereas groups such as the AWS and AWB are principally concerned with issues which relate to women lawyers as professionals, ROW campaigns to improve the law as it affects all women. It is involved in making recommendations to government, lawyers, and other organisations. It publishes research, position papers, newsletters and the *Rights of Women Bulletin* and has a strong media presence. It seeks to inform women of their rights and does this by producing and distributing a range of information leaflets, as well as running a free and confidential legal advice service.

Society of Black Lawyers

The Society of Black Lawyers (SBL) is a national organisation, established to promote the professional needs and goals of African, Asian and Caribbean lawyers, law students, legal workers and legal scholars in the UK. It was founded in 1973 and its primary purpose is to serve as the legal arm for the protection of Britain's black and Asian communities against racism and racial discrimination. The group is involved in campaigning work, publishes regular newsletters and organises evening meetings and events.

Trainee Solicitors Group

The Trainee Solicitors Group (TSG) represents CPE and LPC students, trainees and first year newly-qualified solicitors, all of whom are automatically members. The group, as with the AWS, obtains its income from a grant from the Law Society and via sponsorship. The main aims of the group are to improve the quality of training, encourage a fairer distribution of the financial burdens of legal training, ensure equal access to the profession and develop links with other young trainees and solicitors. Regional groups organise social events and lectures, the annual conference is addressed by high profile speakers and representatives of the TSG sit on a number of Law Society committees. The TSG runs a number of helplines for members with problems or queries. There are, for example, nominated individuals who will deal with queries on finding a training contract, withdrawal of a training contract, debt problems, newly-qualified problems and other problems during a training contract. For details of who to contact, speak to the Law Society.

Young Women Lawyers

Young Women Lawyers (YWL) was formed in 1995 and draws its membership from women under ten year's qualification from all areas of the legal profession, including law students, legal executives, trainee solicitors, pupil barristers, solicitors, barristers and some partners. The group therefore brings together women from all sides of the legal profession and aims to represent the interests of those at the early stages of their training. The group also encourages links between all branches of the profession and campaigns to eradicate discrimination within the legal profession. YWL receives no funding from professional bodies such as the Law Society, and therefore carries out its activities solely with the support of membership fees and donations.

The group organises a wide range of events, focusing on such things as securing a training contract or pupillage, personal development issues such as marketing, mentoring and interviewing, as well as hosting lectures and dinners with prominent women lawyers as speakers. Previous speakers have included Cherie Booth QC, Mrs Justice Hale and Baroness Helena Kennedy QC. The groups produce a newsletter four times a year for members which keeps them up to date with developments in the profession. In addition to the lectures and events, YWL is a campaigning organisation. Each year it has carried out detailed surveys into the position of women in the solicitors' profession which have been covered in detail in the legal and national press and discussed in chapters three and four. YWL believes that these campaigns keep the issue of the role and status of women at the forefront of policy debate within the legal profession.

Appendix 7

List of contributors

Geraldine Andrews	Barrister, Essex Court Chambers
Mrs Justice Arden	High Court Judge, Chairman of the Law Commission
Kamlesh Bahl	Chair, Equal Opportunities Commission, Deputy Vice-President of the Law Society 1998–1999
Fiona Boyle	Senior Lecturer in Law, School of Law, University of Northumbria at Newcastle
Laura Cox	QC, Cloisters
Ruth Deech	Principal, St Anne's College, Oxford
Mary Dowson	Principal Legal Executive, Middlesborough Council
Evelyn Ellis	Professor of Law, Faculty of Law, University of Birmingham
Diana Good	Partner, Linklaters
Katie Goshe	Barrister, 1 Pump Court
Caroline Graham	Solicitor
Josephine Hayes	Barrister, 3 New Square
Mrs Justice Hale	High Court Judge
Joansne Hall	Solicitor
Barbara Hewson	Barrister, Littman Chambers, 12 Gray's Inn Square
Rosalyn Higgins	QC, Judge of the International Court of Justice
Jane Hyndman	Solicitor, Live TV!
Najma Khanzada	Barrister, 1 Pump Court

Sadie King	PhD Student, Newcastle Law School, University of Newcastle upon Tyne
Clare Maher	Case Worker, Legal Aid Board
Gillian Marley	District Judge
Barbara Mills	QC, Director of Public Prosecutions
Katherine O'Donovan	Professor of Law, Faculty of Laws, Queen Mary & Westfield College, University of London
Jo Shaw	Professor of Law, Faculty of Law, University of Leeds
Natalia Siabkin	Solicitor
Hilary Sommerlad	Senior Lecturer in Law, Department of Law, Leeds Metropolitan University
Rachel Tetzlaff-Sarfas	Partner, Thomspons
Elizabeth Woodcraft	Barrister, Tooks Court Chambers

Bibliography

Abel, Richard, *The Legal Profession in England and Wales*, Blackwell: Oxford, 1988.

Abel, Richard and Lewis, Philip, *Lawyers in Society: Comparative Theories*, University of California Press: Berkeley, 1989.

Abella, Rosalie, 'The Concept of an Independent Judiciary', in S Martin and K Mahoney (eds), *Equality and Judicial Neutrality*, Carswell: Toronto, 1987.

Acker, Sandra, 'Women: the other academics', 1 (1980) *British Journal of the Sociology of Education* 81–91.

Adler, Mark, *Clarity for Lawyers: the use of plain English in legal writing*, Law Society: London, 1990.

Aliotta, Jilda, 'Justice O'Connor and the equal protection clause: a feminine voice?', 78 (1995) *Judicature* 232–235.

Armstrong, Nick, 'The training contract under review', 148 (1998) *New Law Journal* 216–218.

Armstrong, Nick and Moorhead, Richard, 'Bare minimum', 147 (1997) *New Law Journal* 487–489.

Asprey, Michele, *Plain Language for Lawyers*, Federation Press: Australia, 1991.

Association of University Teachers, *Sex Discrimination in Universities*, AUT: London, 1992.

Association of University Teachers, *Equal Opportunities, Employment and domestic responsibilities*, AUT: London, 1997.

Association of University Teachers, *Pride not prejudice – equal opportunities in higher education: a lesbian, gay and bisexual perspective*, AUT: London, 1998.

Association of Women Barristers, *Memorandum by the AWB – Judicial Appointments Procedures*, AWB: London, 1995.

Atkins, Susan and Hoggett, Brenda, *Women and the Law*, Blackwell: Oxford, 1984.

Auchmuty, Rosemary, 'Review of Hilaire Barnett, *Sourcebook on Feminist Jurisprudence*', 6 (1998) *Feminist Legal Studies* 135–137.

Australian Law Reform Commission, *Equality before the Law: Women's Equality*, Report No 69, 1994.

Bagilhole, Barbara, 'How to keep a good woman down: an investigation of the role of institutional factors in the process of discrimination against women academics', 14 (1993) *British Journal of the Sociology of Education* 261–274.

Barnett, Hilaire, *Sourcebook on Feminist Jurisprudence*, Cavendish: London, 1997.

Bartlett, Katherine, 'Feminist Legal Methods', 103 (1990) *Harvard Law Review* 829–888.

Barton, Catherine and Farrelly, Catherine, 'Women in the legal profession – a student perspective', 148 (1998) *New Law Journal* 599.

Baty, Phil, '"Elitism" hinders learners', 14 February 1997, *Times Higher Educational Supplement.*

Bean, David, *Law Reform for All*, Blackstone: London, 1996

Becker, Gary, *A Treatise on the Family*, Harvard University Press: Mass, USA, 1991.

Bellerby, Steve and Harris, Phil, *A Survey of Law Teaching 1993*, Sweet & Maxwell and the Association of Law Teachers: London, 1993.

Bindman, Dan, 'Assistants baulk at long hours', *Law Society Gazette*, 13 August 1997.

Birks, Peter, 'Rough justice for law students', *The Times*, 14 September 1993.

Birks, Peter (ed), *Reviewing Legal Education*, Oxford University Press: Oxford, 1994.

Bojarski, Andrzej and Kirk, Jonathan, 'A Modern Route to the Bar', *Counsel*, May/June 1997.

Booth, Cherie, 'Equality: The Bar in the 21st Century', 2 (1998) *Inter Alia* 3–4.

Bottomley, Anne (ed), *Feminist Perspectives on the Foundational Subjects of Law*, Cavendish: London, 1996.

Bottomley, Anne, 'Feminism, the Desire for Theory and the Use of Law', in Hilaire Barnett (ed), *Sourcebook on Feminist Jurisprudence*, Cavendish: London, 1997.

Boyle, Fiona, *Legal Careers*, Fourth Estate and Guardian Careers Guides, 1997.

Bradney, Anthony and Cownie, Fiona, *English Legal System in Context*, Butterworths: London, 1996.

Brill, Laura and Ginsburg, Ruth Bader, 'Women in the Federal Judiciary', 64 (1995) *Fordham Law Review* 281–290.

Briscoe, Ivan and Wilkinson, Helen, *Parental Leave: the price of family values?*, Demos: London, 1997.

Charter, David and O'Leary, John, 'Oxford prepares the way for more women firsts', *The Times*, 29 January 1998.

Canadian Bar Association, *Touchstones for Chance: Equality, Diversity and Change*, 1993.

Clark, Veve et al (eds), *Anti-Feminism in the Academy*, Routledge: London, 1996.

Cole, Bill, *Solicitors in Private Practice: Findings from the Law Society Omnibus Survey*, Law Society: London, 1997.

Cole, Bill, *Trends in the Solicitors' Profession – Annual Statistical Report 1997*, Law Society: London, 1997.

Colker, Ruth, *Pregnant Men – Practice, Theory and the Law*, Indiana University Press: USA, 1994.

Collier, Richard, 'Masculinism, Law and Law Teaching', 19 (1989) *International Journal of the Sociology of Law* 427–451.

Collier, Richard, '"Nutty Professors", "Men in Suits" and "New Entrepreneurs": Corporeality, Subjectivity and Change in the Law School and Legal Practice', 7 (1998) *Social and Legal Studies* 27–53.

Committee of Inquiry into Equal Opportunities on the Bar Vocational Course, Final Report, 1994.

Committee of Vice Chancellors and Principals, *Promoting People: a strategic framework for the management and development of staff in UK universities*, London, 1993.

Cooper, Glenda, 'Lord Irvine and a spot of bother with women', *The Independent*, 21 February 1998.

Cooper, Glenda, 'Female lawyers told dress holds you back', *The Independent*, 2 March 1998.

Cownie, Fiona, 'Women Legal Academics: A New Research Agenda?', 25 (1998) *Journal of Law and Society* 102–115.

Cownie, Fiona and Bradney, Anthony, *English Legal System in Context*, Butterworths: London, 1996.

Cox, Laura and Hewson, Barbara, 'Dealing with Harassment – A Practical Guide', *Counsel*, May/June 1996.

Cox, Laura and Hewson, Barbara, 'Equality of Work Distribution in Chambers', *Counsel*, September/October 1996.

Danusia, Malina and Leonard, Pauline, 'Caught Between Two Worlds: Mothers as Academics', in Sue Davues et al (eds), *Changing the Subject – Women in Higher Education*, Taylor & Francis: London, 1994.

David, Miriam and Woodward, Diana (eds), *Negotiating the Glass Ceiling – careers of senior women in the academic world*, Falmer Press: London, 1998.

Davidson, Gillian and Sidaway, Judith, *The Panel: A Study of Private Practice 1996–97*, Law Society: London, 1998.

Davies, Celia and Holloway, Penny, 'Troubling Transformations: Gender Regimes and Organisational Culture in the Academy', in Louise Morley and Val Walsh (eds), *Feminist Academics: Creative Agents for Change,* Taylor & Francis: London, 1995.

Davues, Sue et al (eds), *Changing the Subject – Women in Higher Education*, Taylor & Francis: London, 1994.

de Beauvoir, Simone, *The Second Sex*, translated by H M Parshley, Penguin Books: London, 1953.

de Groot, Joanna, 'After the Ivory Tower: Gender, commodification and the "academic"', 55 (1997) *Feminist Review* 130–142.

Desole, G and Hoffman, L, (eds), *Rocking the Boat: Academic Women and Academic Processes*, Modern Languages Association of America: New York, 1981.

Doran, Anthony, 'Lawyers hold no brief for equality code', *The Daily Mail*, 7 November 1995.

Dusky, Lorraine, *Still Unequal – The Shameful Truth about Women and Justice in America*, Crown Publishers: New York, 1996.

Dyer, Clare, 'Lord Mackay urges more women to apply for silk', *The Guardian*, 10 April 1995.

Dyer, Clare, 'Women bar students "are regularly offered traineeships for sex"', *The Guardian*, 4 May 1995.

Dyer, Clare, 'Women QCs "enjoy reverse bias"', *The Guardian*, 30 September 1996.

Dyer, Clare, 'Women solicitors "unequal partners" in law firms', *The Guardian*, 8 July 1997.

Eggins, Heather, 'Famous five fail to redress balance', *Times Higher Educational Supplement*, 11 July 1997.

Equal Opportunities Commission, *The Appointments Procedures for Judges and Magistrates*, Memorandum to Home Affairs Committee 1995–96, EOC: Manchester, 1995.

Equal Opportunities Commission, *The Lifecycle of Inequality*, EOC: Manchester, 1995.

Equal Opportunities Commission, 'EOC Response to the National Committee of Inquiry into Higher Education', EOC: Manchester, 1996.

Equal Opportunities Review, 'Parental and family leave', 66 (1996) *Equal Opportunities Review* 22–29.

Farrelly, Catherine and Barton, Catherine, 'Women in the legal profession – a student perspective', 148 (1998) *New Law Journal* 599.

Fineman, Martha Albertson and Thomadsen, Nancy Sweet (eds), *At the Boundaries of Law – Feminism and Legal Theory*, Routledge: London, 1991.

Fitzpatrick, Peter (ed), *Dangerous Supplements*, Pluto Press: London, 1991.

Flood, John, *Barristers' Clerks: the law's middlemen*, Manchester University Press: Manchester, 1993.

Fogarty, Chris, 'Equal rights boss enters Law Soc race', *The Lawyer*, 28 October 1997.

Fox, Marie and Lee, Simon, *Learning Legal Skills*, Blackstone: London, 1994.

Fredman, Sandra, *Women and the Law*, Oxford University Press: Oxford, 1997.

Frug, Mary Joe, 'Re-reading Contracts: A Feminist Analysis of a Contracts Casebook', 34 (1985) *American University Law Review* 1065.

Garza, Hedda, *Barred from the Bar – A History of Women in the Legal Profession*, Franklin Watts: New York, 1996.

Gaudron, Mary, 'Speech to Launch Australian Women Lawyers', 72 (1998) *Australian Law Journal* 119–124.

Gibb, Frances, 'Labour promises to bear down hard on legal aid costs', *The Times*, 30 September 1996.

Gibb, Frances, 'Bar to launch spot checks on compliance with equality code', *The Times*, 12 April 1997.

Gibb, Frances, 'Working mum reaches the top', *The Times*, 6 January 1998.

Gibb, Frances, 'An end to the old sex war?', *The Times*, 3 February 1998.

Gibb, Frances, 'Inquiry into "secret" law jobs', *The Times*, 26 February 1998.

Gibb, Frances, 'Sexual bias is still a problem, says a Bar pupil', *The Times*, 27 April 1998.

Gibson, Susie, 'Define and Empower: Women Students Consider Feminist Learning', 1 (1990) *Law and Critique* 47–60.

Gilligan, Carol, *In a Different Voice – Psychological theory and women's development*, Harvard University Press: Harvard, USA, 1982.

Ginsburg, Ruth Bader and Brill, Laura, 'Women in the Federal Judiciary', 64 (1995) *Fordham Law Review* 281–290.

Gilvarry, Evlynne, 'Outspoken voice strives for top job', *Law Society Gazette*, 12 April 1995.

Gilvarry, Evlynne and Smerin, Jessica, 'Market shy', *Law Society Gazette*, 12 April 1995.

Goldston, James, 'Pregnant Pause', *Legal Business*, 1992.

Gow, Elizabeth, 'Free representation', *Counsel*, November/December 1996.

Graffy, Colleen, 'Coming to terms with keeping terms', *Counsel*, March/April 1997.

Graham, Caroline and McGlynn, Clare, *Soliciting Equality – Equality and Opportunity in the Solicitors' Profession*, Young Women Lawyers: London, 1995.

Graycar, Regina (ed), *Dissenting Opinions – Feminist Explorations in Law and Society*, Allen & Unwin: Sydney, 1990.

Graycar, Regina, 'The Gender of Judgments: An Introduction', in Margaret Thornton, (ed), *Public and Private – Feminist Legal Debates*, Oxford University Press: Oxford, 1995.

Graycar, Regina and Morgan, Jenny, *The Hidden Gender of Law*, Federation Press: Australia, 1990.

Graycar, Regina and Morgan, Jenny, 'Legal Categories, Women's Lives and the Law Curriculum', 18 (1996) *Sydney Law Review* 431–450.

Griffith, JAG, *The Politics of the Judiciary*, 5th ed, Fontana Press: London, 1997.

Griffiths, Sian (ed), *Beyond the Glass Ceiling – Forty women whose ideas shape the modern world*, Manchester University Press: Manchester, 1996.

Griffiths, Sian, 'Chipping away at the glass ceiling', *Times Higher Educational Supplement*, 26 July 1996.

Griffiths, Sian, 'The struggle for equality', *Times Higher Educational Supplement*, 6 June 1997.

Gubbay, Joanne, 'My learned friends want to be more flexible', *The Times*, 14 April 1998.

Guinier, Lani et al, 'Becoming Gentlemen: Women's Experiences at One Ivy League Law School', 143 (1994) *University of Pennsylvania Law Review* 1-110.

Hagan, John and Kay, Fiona, *Gender in Practice – A Study of Lawyers' Lives*, Oxford University Press: Oxford, 1995.

Hansard Society, *Women at the Top*, Hansard Society: London, 1989.

Harriet Swain, 'First-class degrees on wane', *Times Higher Educational Supplement*, 4 July 1997.

Harrington, Mona, *Women Lawyers – Rewriting the Rules*, Plume-Penguin: USA, 1995.

Harris, Phil and Bellerby, Steve, *A Survey of Law Teaching 1993*, Sweet & Maxwell and the Association of Law Teachers: London, 1993.

Harris, Phil and Jones, Martin, 'A Survey of Law Schools in the UK 1996', 31 (1997) *Law Teacher* 38–126.

Hayes, Josephine, 'Appointment by invitation', 147 (1997) *New Law Journal* 520–522.

Heward, Christine, 'Academic Snakes and Ladders: reconceptualising the "glass ceiling"', 6 (1994) *Gender and Education* 249–262.

Hewson, Barbara, 'Sexual Harassment at the Bar – A recent problem?', 145 (1995) *New Law Journal* 626–627.

Hewson, Barbara, 'Why women get a raw deal in the courts', *The Times*, 17 September 1996.

Hewson, Barbara, 'You've a long way to go, baby ...', 146 (1996) *New Law Journal* 565–566.

Hewson, Barbara, and Cox, Laura, 'Dealing with Harassment – A Practical Guide', *Counsel*, May/June 1996.

Hewson, Barbara, and Cox, Laura, 'Equality of Work Distribution in Chambers', *Counsel*, September/October 1996.

Hilder, Paul, *IT in the Solicitors' Office*, Blackstone Press: London, 1997.

Hoffman, L and Desole, G (eds), *Rocking the Boat: Academic Women and Academic Processes*, Modern Languages Association of America: New York, 1981.

Hoggett, Brenda and Atkins, Susan, *Women and the Law*, Blackwell: Oxford, 1984.

Holloway, Penny and Davies, Celia, 'Troubling Transformations: Gender Regimes and Organizational Culture in the Academy', in Louise Morley and Val Walsh (eds), *Feminist Academics: Creative Agents for Change,* Taylor & Francis: London, 1995.

Home Affairs Committee, Third Report of Session 1995–6, Volume I, *Judicial Appointments Procedures*.

Home Affairs Committee, Third Report of Session 1995–6, Volume II, *Minutes of Evidence and Appendices*.

Home Affairs Committee, Third Report of Session 1996–7, *Freemasonry in the Police and the Judiciary*.

Hughes, Sally, *The Circuit Bench – A Woman's Place?*, Law Society: London, 1991.

Irvine, Lord, 'The Legal System and Law Reform under Labour', in David Bean (ed), *Law Reform for All*, Blackstone: London, 1996

Jack, Dana Crowley and Jack, Rand, *Moral Vision and Professional Decisions: The Changing Values of Women and Men Lawyers*, Cambridge University Press: New York, 1989.

Jack, Rand and Jack, Dana Crowley, *Moral Vision and Professional Decisions: The Changing Values of Women and Men Lawyers*, Cambridge University Press: New York, 1989.

Jenkins, John and Walker, Danielle, *Annual Statistical Report 1993*, Law Society: London, 1993.

Jones, Martin and Harris, Phil, 'A Survey of Law Schools in the UK 1996', 31 (1997) *Law Teacher* 38–126.

Jordan, Emma Coleman, 'Images of Black Women in the Legal Academy: An Introduction', 6 (1990–91) *Berkeley Women's Law Journal* 1–21.

Judd, Judith, 'Bias that stops women academics reaching the top', *The Independent*, 7 June 1997.

Justice, *The Judiciary in England and Wales*, Justice: London, 1992.

Kairys, David (ed), *The Politics of Law*, Pantheon Books: New York, 1990.

Kay, Fiona and Hagan, John, *Gender in Practice – A Study of Lawyers' Lives*, Oxford University Press: Oxford, 1995.

Kennedy, Duncan, 'Legal Education as Training for Hierarchy', in David Kairys (ed), *The Politics of Law*, Pantheon Books: New York, 1990.

Kennedy, Helena, 'Women at the Bar', in Robert Hazell (ed), *The Bar on Trial*, Quartet Books: London, 1978.

Kennedy, Helena, *Eve was Framed – Women and British Justice*, Chatto & Windus: London, 1992.

Helena Kennedy, 'Introduction', in Sian Griffiths (ed), *Beyond the Glass Ceiling – Forty women whose ideas shape the modern world*, Manchester University Press: Manchester, 1996, pp 1–9.

Kennedy, Helena, 'Who's been sitting in my chair?', *The Guardian*, 8 July 1996.

Kenny, Phillip, *Studying Law*, 4th ed, Butterworths, 1998.

Killick, Judith, and Pigott, Maggy, 'For the record', *Counsel,* September/October 1997.

Kirk, Harry, *Portrait of a Profession*, Oyez: London, 1976.

Kirk, Jonathan and Bojarski, Andrzej, 'A Modern Route to the Bar', *Counsel*, May/June 1997.

Laband, David and Lentz, Bernard, *Sex Discrimination in the Legal Profession*, Quorum Books: USA, 1995.

Law Society, *Equal in the Law*, Law Society: London, 1988.

Law Society, *Trends in the Solicitors' Profession – Annual Statistical Report 1994*, Law Society: London, 1995.

Lee, Robert, 'A Survey of Law School Admissions', 18 (1984) *The Law Teacher* 165.

Lee, Simon and Fox, Marie, *Learning Legal Skills*, Blackstone: London, 1994.

Leighton, Patricia, Mortimer, Tom and Whatley, Nicola, *Today's Law Teachers: Lawyers or Academics?*, Cavendish: London, 1995.

Lentz, Bernard and Laband, David, *Sex Discrimination in the Legal Profession*, Quorum Books: USA, 1995.

Leonard, Pauline and Malina, Danusia, 'Caught Between Two Worlds: Mothers as Academics', in Sue Davues et al (eds), *Changing the Subject – Women in Higher Education*, Taylor & Francis: London, 1994.

Levin, Michael, 'Women, Work, Biology, Justice', in Caroline Quest (ed), *Equal Opportunities: A Feminist Fallacy*, Institute of Economic Affairs: London, 1992.

Lewis, Philip and Abel, Richard, *Lawyers in Society: Comparative Theories*, University of California Press: Berkeley, 1989.

Lewis, Verity, *Trends in the Solicitors' Profession – Annual Statistical Report 1996*, Law Society: London, 1996.

Lieberman, Marcia, 'The most important thing for you to know', in G Desole and L Hoffman (eds), *Rocking the Boat: Academic Women and Academic Processes,*, Modern Languages Association of America: New York, 1981.

Lindsay, Robert, 'You really wouldn't expect it of law firms', *The Lawyer*, 28 January 1997.

Longrigg, Clare, 'If you want to get ahead then get a suit, women lawyers told', *The Guardian*, 2 March 1998.

Lord Chancellor's Advisory Committee on Legal Education and Conduct, *First Report on Legal Education and Training*, ACLEC: London, 1996.

Loux, Andrea, 'Idols and Icons: Mackinnon Catherine and Freedom of Expression in North America', 6 (1998) *Feminist Legal Studies* 85–104.

Lyon, Kate and West, Jackie, 'The Trouble with Equal Opportunities: the case of women academics', 7 (1995) *Gender and Education* 51–68.

MacLeod, Donald, 'Confidence Trick', *The Guardian*, 13 January 1998.

Martin, Robyn, 'A Feminist View of the Reasonable Man', 25 (1994) *Anglo-American Law Review* 334–374.

Malleson, Kate, 'The Use of Judicial Appointments Commissions: A Review of the US and Canadian Models', Research Series No 6/97, LCD, 1997.

Mansell, Wade et al, *A Critical Introduction to Law*, Cavendish: London, 1995.

McCann, Paul, 'Rumpole sentenced to an early retirement', *The Independent*, 13 January 1998.

McGlynn, Clare 'Soliciting Equality – the way forward', 145 (1995) *New Law Journal* 1065–1066, 1070.

McGlynn, Clare, 'Sex Discrimination at the Margins', 146 (1996) *New Law Journal* 379–381.

McGlynn, Clare, 'Paying for Equality', 147 (1997) *New Law Journal* 568–569.

McGlynn, Clare, 'Where men still rule', *The Times*, 22 April 1997.

McGlynn, Clare, 'Time is ripe for parental leave', *The Times*, 27 May 1997.

McGlynn, Clare, 'The Business of Equality', in Clare McGlynn (ed), *Legal Feminisms: theory and practice*, Dartmouth: Aldershot, 1998.

McGlynn, Clare (ed), *Legal Feminisms: theory and practice*, Dartmouth: Aldershot, 1998.

McGlynn, Clare and Graham, Caroline, *Soliciting Equality – Equality and Opportunity in the Solicitors' Profession*, Young Women Lawyers: London, 1995,

McNabb, Robert and Wass, Victoria, 'Male–female Salary Differentials in British Universities', 49 (1997) *Oxford Economic Papers* 328.

McNay, Ian, 'The Impact of the 1992 RAE on Individual and Institutional Behaviour in English Higher Education', Centre for Higher Education Management, Anglia Polytechnic University, 1997.

McRae, Susan, *Women at the Top – Progress After Five Years*, Hansard Society: London, 1996.

Mears, Martin, 'Aux armes to defend the revolution', *The Times*, 16 April 1996.

Melling, Louise and Weiss, Catherine, 'The Legal Education of Twenty Women', 40 (1988) *Stanford Law Review* 1299–1369.

Menkel-Meadow, Carrie, 'Portia in a different voice: speculations on a woman's lawyering process', 1 (1985) *Berkeley Women's Law Journal* 39–63.

Menkel-Meadow, Carrie, 'Feminisation of the Legal Profession: the comparative sociology of women lawyers', in Richard Abel and Philip Lewis, *Lawyers in Society: Corporate Theories*, University of California Press: Berkeley, 1989.

Millsom, Henry, 'Ivory Towers and Ebony Women: the experiences of black women in higher education', in Sue Davies et al (eds), *Changing the Subject – Women in Higher Education*, Taylor & Francis: London, 1994.

Moorhead, Richard and Armstrong, Nick, 'Bare minimum', 147 (1997) *New Law Journal* 487–489.

Morgan, Jenny and Graycar, Regina, *The Hidden Gender of Law*, The Federation Press: Australia, 1990.

Morgan, Jenny and Graycar, Regina, 'Legal Categories, Women's Lives and the Law Curriculum', 18 (1996) *Sydney Law Review* 431–450.

Morley, Louise, 'Glass Ceiling or Iron Cage: Women in UK Academia', 1 (1994) *Gender, Work and Organization* 194.

Morley, Louise and Walsh, Val, (eds), *Feminist Academics: Creative Agents for Change*, Taylor & Francis: London, 1995.

Morley, Louise and Walsh, Val, 'Feminist Academics: Creative Agents for Change', in Louise Morley and Val Walsh (eds), *Feminist Academics: Creative Agents for Change*, Taylor & Francis: London, 1995.

Morris, Gillian and Wilson, William, 'The Future of the Academic Law Degree', in Peter Birks (ed), *Reviewing Legal Education*, Oxford University Press: Oxford, 1994.

Morrison, Mairi, 'May it Please the Court?: How moot court perpetuates gender bias in the "real world" of practice', 6 (1995) *UCLA Women's Law Journal* 49–84.

Mortimer, Tom, Whatley, Nicola and Leighton, Patricia, *Today's Law Teachers: Lawyers or Academics?*, Cavendish: London, 1995.

Mossman, Mary Jane, 'Otherness and the Law School: A Comment on Teaching Gender Equality', 1 (1985) *Canadian Journal of Women and the Law* 213.

Mossman, Mary Jane, 'Women Lawyers in twentieth century Canada: rethinking the image of Portia', in Regina Graycar (ed), *Dissenting Opinions – Feminist Explorations in Law and Society*, Allen & Unwin: Sydney, 1990.

Mossman, Mary Jane, 'Feminism and Legal Method: The Difference it Makes', in Martha Albertson Fineman and Nancy Sweet Thomadsen (eds), *At the Boundaries of Law – Feminism and Legal Theory*, Routledge: London, 1991.

Mossman, Mary Jane, 'The Use of Non-Discriminatory Language in the Law', 73 (1994) *Canadian Bar Review* 347–371.

Naylor, Bronwyn, 'Pregnant Tribunals', 14 (1989) *Legal Service Bulletin* 41.

Newburn, Tim and Shiner, Michael, *Entry into the Legal Professions – The Law Student Cohort Study Year 3*, Law Society: London, 1995.

Nott, Sue, 'Women in the Law', 139 (1989) *New Law Journal* 749–750.

Nussbaum, Martha, *Cultivating Humanity – A Classical Defense of Reform in Liberal Education*, Harvard University Press: Harvard, 1997.

O'Connor, Sandra Day, 'Portia's Progress', 66 (1991) *New York University Law Review* 1546–1558.

O'Leary, John, 'University girls fall behind in confidence', *The Times*, 5 January 1998.

Oliver, Dawn, 'Teaching and Learning Law: Pressures on the Liberal Law Degree', in Peter Birks (ed), *Reviewing Legal Education*, Oxford University Press: Oxford, 1994.

Palmer, Camilla, *Maternity Rights*, Legal Action Group: London, 1996.

Payne, Jennifer, 'Limiting the liability of professional partnerships: in search of this Holy Grail', 18 (1997) *Company Lawyer* 81–88.

Peers, Ian, 'Gender and age bias in the predictor-criterion relationship of A levels and degree performance: a logistic regression analysis', (1994) *Research in Education No 52*.

Phillips, Estelle and Pugh, Derek, *How to get a PhD*, 2nd ed, Open University Press: Buckingham, 1996.

Pigott, Maggy and Killick, Judith, 'For the record', September/October (1997) *Counsel* 16–19.

Podmore, David and Spencer, Anne, 'Women Lawyers in England', 9 (1982) *Work and Occupations* 337–361.

Pugh, Derek and Phillips, Estelle, *How to get a PhD*, 2nd ed, Open University Press: Buckingham, 1996.

Quest, Caroline (ed), *Equal Opportunities: A Feminist Fallacy*, Institute of Economic Affairs: London, 1992.

Reeves, Peter, *Silk Cut: Are QCs really necessary?*, Adam Smith Institute: London, 1998.

Rice, Robert, 'Female attraction', *The Financial Times*, 3 June 1997.

Rhode, Deborah, 'Perspectives on Professional Women', 40 (1988) *Stanford Law Review* 1163–1207.

Rose, Neil, 'In-house teams in growth boom', *Law Society Gazette*, 9 July 1997.

Rose, Neil, 'TSG moots mixing work with study', *Law Society Gazette*, 28 January 1998.

Rose, Neil, 'In-house teams "attract women"', *Law Society Gazette*, 4 February 1998.

Royal Commission on Legal Services, Cmnd 7648, HMSO: London, 1979.

Rozenburg, Joshua, *The Search for Justice*, Sceptre: London, 1994.

Sanderson, Peter and Sommerlad, Hilary, *Gender, Choice and Commitment: a study of women lawyers*, Dartmouth-Ashgate: Aldershot, forthcoming.

Schneider, Elizabeth, 'Task Force Reports on Women in the Courts': the challenge for legal education', 38 (1998) *Journal of Legal Education* 87–99.

Sedley, Stephen, 'Human Rights: a twenty-first century agenda', (1995) *Public Law* 386.

Shapland, Joanna and Sorsby, Angela, *Starting Practice: work and training at the junior Bar*, Institute for the Study of the Legal Profession: University of Sheffield, 1995.

Sherry, Suzanna, 'Civil virtue and the feminine voice of constitutional adjudication', 72 (1986) *Vanderbilt Law Review* 543–615.

Sherry, Suzanna, 'The Gender of Judges', 4 (1986) *Law and Inequality* 159.

Shiner, Michael, *Entry into the Legal Professions – the Law Student Cohort Study Year 4*, Law Society: London, 1997.

Shiner, Michael and Newburn, Tim, *Entry into the Legal Professions – The Law Student Cohort Study Year 3*, Law Society: London, 1995.

Shuaib, S, 'Discrimination still exists', *The Independent*, 21 June 1991.

Sidaway, Judith and Davidson, Gillian, *The Panel: A Study of Private Practice 1996–97*, Law Society: London 1998.

Skordaki, Eleni, 'Glass slippers and glass ceilings: women in the legal profession', 3 (1996) *International Journal of the Legal Profession* 7–43.

Smart, Carol, 'Feminist Jurisprudence', in Peter Fitzpatrick (ed), *Dangerous Supplements*, Pluto Press: London, 1991.

Smerin, Jessica, 'Brothers in law', *Law Society Gazette*, 5 February 1997.

Smerin, Jessica and Gilvarry, Evlynne, 'Market shy', *Law Society Gazette*, 12 April 1995.

Sommerlad, Hilary, 'The myth of feminisation: women and cultural change in the legal profession', 1 (1994) *International Journal of the Legal Profession* 31–53.

Sommerlad, Hilary, 'The Gendering of the Professional Subject: commitment, choice and social closure in the Legal Profession', in Clare McGlynn (ed), *Legal Feminisms: theory and practice*, Dartmouth-Ashgate: Aldershot, 1998.

Sommerlad, Hilary and Sanderson, Peter, *Gender, Choice and Commitment: a study of women lawyers,* Dartmouth-Ashgate: Aldershot, 1998.

Spencer, Anne and Podmore, David, 'Women Lawyers in England', 9 (1982) *Work and Occupations* 337–361.

Stapley, Sue, 'Social Climbing', *Law Society Gazette*, 25 September 1996.

Swain, Harriet, 'Minorities must not be ignored', *Times Higher Educational Supplement*, 13 June 1997.

Symposium, 'Black Women Law Professors: Building a Community at the Intersection of Race and Gender', 6 (1990–91) *Berkeley Women's Law Journal* 1–201.

Teitelbaum, Lee et al, 'Gender, Legal Education and Legal Careers', 41 (1991) *Journal of Legal Education* 443–481.

Thomadsen, Nancy Sweet and Fineman, Martha Albertson (eds), *At the Boundaries of Law – Feminism and Legal Theory*, Routledge: London, 1991.

Thomas, Cheryl, 'Judicial Appointments in Continental Europe', Research Series No 6/97, LCD, 1997.

Thornton, Margaret, 'Hegemonic masculinity and the academy', 17 (1989) *International Journal of the Sociology of Law* 115–130.

Thornton, Margaret, 'Discord in the Legal Academy: the case of the feminist scholar', 3 (1994) *Australian Feminist Law Journal* 53.

Thornton, Margaret, (ed), *Public and Private – Feminist Legal Debates*, Oxford University Press: Oxford, 1995.

Thornton, Margaret, *Dissonance and Distrust – Women in the Legal Profession*, Oxford University Press: Oxford, 1996.

TMS Consultants, *Without Prejudice? Sex Equality at the Bar and in the Judiciary*, Bar Council and LCD, 1992.

Towers, Rebecca, 'Sole practice enables many women to combine work and family', *Law Society Gazette*, 16 April 1998.

Trapp, Roger, 'Quarter of female solicitors harassed', *The Independent*, 27 January 1997.

Tzannes, Maria, 'Strategies for the Selection of Students to Law Courses in the 21st Century: Issues and Options for Admissions Policy Makers', 29 (1995) *Law Teacher* 43–63.

UCAS, *Annual Report 1995 Entry*, UCAS: Cheltenham, 1996.

United Nations, Blue Books Series Vol VI, *The Advancement of Women 1945–1995*, UN: New York, 1995.

Walsh, Val and Morley, Louise, (eds), *Feminist Academics: Creative Agents for Change*, Taylor & Francis: London, 1995.

Walsh, Val and Morley, Louise, 'Feminist Academics: Creative Agents for Change', in Louise Morley and Val Walsh (eds), *Feminist Academics: Creative Agents for Change*, Taylor & Francis: London, 1995.

Walker, Danielle and Jenkins, John, *Annual Statistical Report 1993*, Law Society: London, 1993.

Wass, Victoria and McNabb, Robert, 'Male-female Salary Differentials in British Universities', 49 (1997) *Oxford Economic Papers* 328.

Waterlows, *Solicitors' and Barristers' Directory*, 1996.

Weiss, Catherine and Melling, Louise, 'The Legal Education of Twenty Women', 40 (1988) *Stanford Law Review* 1299–1369.

West, Jackie and Lyon, Kate, 'The Trouble with Equal Opportunities: the case of women academics', 7 (1995) *Gender and Education* 51–68.

Whatley, Nicola, Mortimer, Tom and Leighton, Patricia, *Today's Law Teachers: Lawyers or Academics?*, Cavendish: London, 1995.

Wilkinson, Helen and Briscoe, Ivan, *Parental Leave: the price of family values?*, London: Demos, 1997.

Willets, Jayne, 'Getting in the business', *Law Society Gazette*, 26 March 1997.

Williams, Nicola, *Without Prejudice*, Headline Feature: London, 1997.

Wilson, Bertha, 'Will women judges really make a difference?', 28 (1990) *Osgoode Hall Law Journal* 507–522.

Wilson, John, 'A Survey of Legal Education in the UK', 9 (1996–97) *Journal of the Society of Public Teachers of Law* 1.

Wilson, John, 'A third survey of university legal education in the United Kingdom', 13 (1993) *Legal Studies* 143–182.

Wilson, William and Morris, Gillian, 'The Future of the Academic Law Degree', in Peter Birks (ed), *Reviewing Legal Education*, Oxford University Press: Oxford, 1994.

Winskell, Lucy, 'Council Impressions', Young Solicitors Group Newsletter, Spring 1997.

Woodward, Diana and David, Miriam (eds), *Negotiating the Glass Ceiling – careers of senior women in the academic world*, Falmer Press: London, 1998.

Wynn Davies, Patricia, 'Prospects for black barristers get worse', *The Independent,* 13 March 1997.

Young, Lola, 'The colour of ivory towers', *Times Higher Educational Supplement*, 5 June 1998.

Index